Ordering the World

Atoms and Elements
Natural Science Books in English, 1600–1900
Sources for the History of Science, 1660–1914
The Nature of Science
Zoological Illustration
The Transcendental Part of Chemistry

ORDERING
THE WORLD

A History of
Classifying Man

David Knight

BURNETT BOOKS
in association with ANDRE DEUTSCH

First published 1981 by Burnett Books Limited
in association with André Deutsch Limited
105 Great Russell Street London WC1
Copyright © 1981 by David Knight
Photoset, printed and bound in Great Britain by
REDWOOD BURN LIMITED
Trowbridge & Esher

ISBN 0 233 97293 5

*For Frances, much of whose life
has been dominated by it.*

Contents

————

List of Illustrations viii

Acknowledgements xi

1 A Fundamental Preoccupation 13

2 An Objective System 36

3 The Artificial System 58

4 The Shape of Nature 82

5 Accordance with Nature 107

6 Everything in its place 131

7 Development 156

8 Epilogue 184

Suggested Further Reading 207

Index 210

List of Illustrations

———

FIGURE 1 The monkeys arranged according to Swainson's Quinary System 96

FIGURE 2 Mendeleev's first Periodic Table 138

FIGURE 3 Whewell's arrangement of the sciences 144

FIGURE 4 Diagram illustrating the appearance, abundance and disappearance of reptile groups 157

FIGURE 5 A cartoon by de la Beche illustrating Lyell's notion that the dinosaurs might return 162

FIGURE 6 Diagram from the *Origin of Species* to illustrate divergence and extinction 164

FIGURE 7 Table showing the evidence for evolution 165

FIGURE 8 William Crookes' arrangement of the Periodic Table 186

FIGURE 9 An evolutionary tree for the parrots 191

We have been led to conclude that, independently of the system of parts marvellously combined to form the individual animal, another more comprehensive system exists, which embraces all animals.

Sir Charles Bell, 1833.

Thus in travelling from one end to the other of the scale of life, we are taught one lesson, that living nature is not a mechanism but a poem; not a mere rough engine-house for the due keeping of pain and pleasure machines, but a palace whose foundations, indeed, are laid on the strictest and safest mechanical principles, but whose superstructure is a manifestation of the highest and noblest art.

Thomas Henry Huxley, 1856.

Acknowledgements

———

I WOULD LIKE to thank the University of Durham for granting me
Sabbatical Leave for the summer of 1979, when the bulk of this
book was written; and for giving me a grant from the Research
Fund to go to London. I would like also to thank those in the
University Library here in Durham for their assistance. I have
made considerable use of the library of the Royal Institution in
London, for which I am grateful; and I have also been welcomed
and assisted by Gavin Bridson, Librarian of the Linnean Society,
who gave general advice and helped me find my way round the
MacLeay and Swainson archives there.

I am also grateful to Mr R. Kitching for taking the photographs
from which the Figures in the text are reproduced.

1

A Fundamental Preoccupation

T HIS IS A BOOK about classifying man. That can be taken in two ways: a discussion of man the classifying animal, anxious to find some order in nature (or if necessary to impose an order there), and also a discussion of how man has fitted himself into this order. We are therefore concerned with the very fundamental questions of how mankind has perceived the world, and how it has perceived itself in that world. These are questions of more than arid or rarified scientific interest, for upon the answers given to the questions, what sort of a world is this? and what is my place in it? have depended, depend and will depend our thinking about the world and its resources, our belief or disbelief in a Creator, and our attitude to our fellow men, especially those of different races. It was because of these serious implications that Charles Darwin's contemporaries, within and without the sciences, took his idea of development through natural selection so seriously. One can write the history of biology in the nineteenth century almost without referring to Darwin, and this has indeed become something of a fashion. Darwin never held academic posts, or worked in a splendidly-equipped laboratory, or attended international congresses, or placed his students in influential professorships: the milestones along the road to a successful career in science which his contemporaries were beginning to erect as science became a profession. But because he raised the fundamental questions and gave rather new answers to them, we can call the nineteenth century 'Darwin's century' as we would not call it, for example, 'Liebig's century'.

It may seem to be one thing to reform the world-view of succeeding generations, like Darwin, and another to be merely pigeon-holing and naming animals and plants very hard to tell apart, as the natural historian (or stamp-collector!) is often supposed to do.

On the other hand Darwin devoted many years of his maturity to the classification of barnacles, on which he wrote the standard work. This was sound tactically, for when the *Origin of Species* came out in 1859, eight years after the first volume on barnacles, nobody could say that it was the work of a mere armchair naturalist. But the barnacles were also interesting in their own right; until 1830 eminent naturalists had always placed them among the molluscs because they looked like a kind of shellfish. The naval surgeon J.V.Thompson then showed that young barnacles were mobile, rather shrimp-like little creatures, and that it was only in middle age that barnacles settled down and fixed themselves to a log or ship. Darwin had studied some curious barnacles on *HMS Beagle* and he could not but be interested by a group in which the adults seemed lower in the scale of nature than the larvae, in which some males seemed to have degenerated to 'mere bags of spermatozoa', and in which the very small gradations that separate species were apparent, Nature making no jumps. Though he did find it a grind, to a Darwin, there is no such thing as mere classification; and without the careful study and ordering of organisms his work would not have possessed that blend of general principles and details that distinguishes really great science – or really great history or literary criticism for that matter.

Again, it may seem that while experience of classifying obscure creatures might have been necessary for Darwin, not only in propagating his theory, but also in formulating it, his work has made it unnecessary for anybody later to do the same sort of thing. Taxonomy (as the science of classification is called) might have become obsolete, like long-division in the electronic era, and while one might look with reverence and awe at the portraits of those who did it one would not actually read their writings or expect to find that they had had anything to say that had any bearing upon our lives. In fact, Darwin's theory did not put an end to taxonomy. He argued that the various 'families' were really families, so that the closer species were in the system the closer was their relationship. But animals and plants do not come bearing their family trees with them, and do not even have surnames to make it easier; relationships have to be inferred from resemblances. What has happened in the last hundred years is that more resemblances have been taken into account, so that creatures can be more precisely placed in order, just as the study of larvae enabled Thompson and Darwin to place the barnacles near their real cousins, the crustaceans.

Classifying man classifies himself too, and it might be supposed that here Darwin's work really marked a new beginning, with his cryptic remark – one of the great understatements – that 'Light will be thrown on the origin of man and his history.' At the beginning of his entertaining work, *The Selfish Gene*, Richard Dawkins quotes with approval an eminent zoologist who believed that all attempts before 1859 to answer such questions as 'What is man?' are worthless, and that we would be better off if we ignored them completely. It is one of the measures of a really great work in science that it makes all its predecessors obsolete. This seems, indeed, to be something which distinguishes science from other intellectual activities, for it is less generally true in the arts; but perhaps we ought to ask why we might want to turn to older answers, and thus move eminent contemporaries to redirect us to more recent works.

If our interest is in current anthropology, psychology or zoology, and especially if we have an examination to pass, then naturally it is the most recent works that are likely to be of most use. They will, implicitely or explicitely, convey a world view, and suggest lines of enquiry methods of analysis and interpretations that have proved to be fruitful. In terms of Thomas Kuhn's philosphy of science, they will present the modern paradigm. The word means 'example', and the great men of the last generation or so and their work is an example to beginners of how to go about the science; what the great men did should be copied, developed, and extended, and what they didn't do should be left alone. So the student of modern science must get up the dogma, and ignore past heretics, or those less offensive figures who just happen to be born before the present dispensation and inhabit more comfortable circles of Hell. Perhaps in the same way, art students might be well-advised not to spend much time studying Byzantine icons, and would-be poets not to immerse themselves overmuch in Chaucer; but for anybody who does not just want to submerge himself in the technicalities of science or art, historical perspective, leading to the realisation that alternative ways of looking at things are possible, must be valuable.

To somebody whose primary interest is not in the currently fashionable paradigms of various sciences, but is in the great general questions of how the world is ordered and where we fit into that order, the years before 1859 are in many ways much more interesting than those which come after it. With the establishment of Darwin's paradigm, questions about the origin of man became a part of what Kuhn calls 'normal science' – whereas what Darwin did was 'revolutionary science'. Normal science is concerned with questions difficult to answer, and revolutionary science with those difficult to ask. The revolutionary idea was that we were cousins of the apes – since then the questions have been the more normal ones, concerned with whether a particular fossil hominid is really in the line of evolution leading to us, just how long ago our lineage diverged from that of the gorillas, and so on. These are undoubtedly interesting questions, but they are not especially fundamental to anybody's understanding of the world. Since this is a book about ordering the world and not a popularisation of a branch of science, it will be mostly concerned with how man's place in nature was perceived in the years before the general acceptance of Darwin's theory, and the twentieth century will be relegated to an epilogue. But one might hope that it will help anybody interested in current work to know something of what lies behind it. Men of science, like the citizens of a country, may choose to ignore their history, but they cannot wish it away; and whether they like it or not, their past has given form to their present. There is a particular irony in Darwinians denying the value of history, since what Darwin did was to take into science a way of reasoning invented by historians.

The story of classifying man will therefore be concentrated on the eighteenth and nineteenth centuries, the period in which he was at his most active and when what physicists now sometimes disparage as 'merely descriptive science' enjoyed its highest prestige. Since a part of the argument is that classifying is a fundamental and very significant human activity, we should begin with a general examination of this activity before we come to follow it in history, and in particular in the history of the life sciences.

The world into which we are born is a booming buzzing confusion, and we only slowly learn to sort out things of like kind. Instinctively in babyhood, and later more self-consciously, we group things together and attach general terms to them; so that instead of a chaos of endless particular things without apparent order, we come to perceive a world with a finite number of classes of things. We thus begin to feel at home, even though the classes may need to be revised (sometimes painfully) and certainly seem to cut across each other so that everything belongs in more than one. We distinguish parts of ourselves from other things, and we then separate things that are accessible from things that are not, like the Moon, for which there is little point in crying. Some classes have sharp lines, and others have fuzzy ones; the divisions between colours, for example, seem hard to learn and are mastered some time after children have got size relations straight, and differ between cultures. Great scientists are Peter Pans, still anxious to classify and explain at an age when most people are concerned with money, power, and sex.

In using general terms at all we are classifying, and we come up straight away against what philosophers call the problem of universals. The problem is whether all members of a group share some essential property, or whether they are just grouped by convention or for convenience. For realists, they share a common essence (if we have got our classification right), while for nominalists general terms are just names of more or less usefulness. The dispute goes back to the ancient Greeks, with Plato being a notable realist, and was of great importance in medieval Europe. If we talk abou red objects, we might be prepared to admit readily enough that 'red' covers many shades, that under different light conditions we might judge differently whether something was red (searching for a red car in a car park lit by sodium lamps is an odd experience, because it looks black), and that there would be borderline cases where one might as truly describe something as orange or as red. So we might not feel disposed to fight very hard for some essence in which all red things shared, and agree that 'red' was no more than a convenient adjective.

If we come to science, things perhaps look a bit different. The class of prime numbers is quite unlike that of red things. We can rigorously define a prime number (it is not divisible by any number), we can prove that there is an infinite number of them, and there are no borderline cases; in whatever light we examine 7 it is prime, while 8 is not. We can thus define the essential property of prime numbers, and we are not dealing with a merely conventional label. On the other hand, we are not dealing with things either, but with abstractions; we do not meet numbers in the world, we meet things. When these are classified, for example into metals and non-metals, the result looks about as messy as it did with redness.

Iron, copper, silver and gold are obviously metals. Chlorine, sulphur, and nitrogen are, equally clearly not metals. Mercury is a bit odd, but even in Antiquity it was accepted as a metal; on the Arctic expeditions of the early nineteenth century experiments were sometimes done on solid mercury, which was otherwise unobtainable and which behaved pretty much like other metals. In 1800, it was one of the essential properties of metals to be heavy, or more strictly, dense. When in 1807 Humphry Davy isolated the 'basis' of potash using an electric current, he found that it floated on water, decomposing it so violently that the hydrogen given off was set on fire. This made a splendid demonstration-experiment to show to the fashionable audiences who flocked to his lectures at the Royal Institution; but it also posed a taxonomic problem, and Davy at first called the new substance by the neutral term 'potagen'. Only when, after a few more days' work, he became convinced that the bulk of its analogies were with the metals did he rename it 'potasium' and then 'potassium', a form indicating metallic status. The status of other substances remained problematic; Davy's 'silicium' has now become 'silicon' to bring out its non-metallic status and relationship to carbon, but elements like arsenic, antimony, and selenium are genuine borderline cases.

This might seem to indicate that the classification of elements into metals and non-metals is not somehow fundamental, but that like 'red' the adjective 'metallic' is sometimes a convenient label and sometimes not, though the tests for whether something is a metal are more sophisticated than those for telling whether it is red. The controversy between nominalists and realists became important when they were discussing goodness rather than redness. We speak of good actions and good men, but also of good poems and pictures, and of good cars. Is there any essential quality common to all these? If there is not, then perhaps 'good' just corresponds to some seal of approval given by the speaker or writer. There may be borderline cases, as where a teacher is unsure whether to mark an essay 'good' merely 'satisfactory'; and when we are uncertain whether Henry VIII was a bad king but a 'good thing' or the other way around. If 'good' is simply a term of approval, then the sage and the sadist will see actions in different lights and apply the term differently; but there will be no universally-agreed definition as there is for 'prime number' and moral judgements will not be absolute.

Moral relativism, like apes for ancestors, is less alarming to us less than it was to those who lived in previous centuries; but even so most people would be unhappy at the thought that goodness was simply a matter of taste or convention, and wrong or evil merely deviance. Nevertheless, in ethics, and inorganic chemistry, the nominalists seem in our day to be on top; only where one can rigorously define classes with a single character can one readily maintain realism. This has sometimes been done in natural history, particularly for the broader groups such as genera; but usually natural historians have, in their search for the natural system, adopted a more flexible approach based upon not one but a series of criteria. This is what Davy was using when he put potassium among the metals. There was an accepted list of metallic properties; and when potassium turned out to have most of them, then he put it in that class.

This kind of classifying goes back to Aristotle, who in his extensive biological writings criticised Plato for forming classes based upon the presence or absence of one character. Plato's man, the 'featherless biped', was the result of such a system and got one to man in two steps; but while it might do as a definition for man it fails to bring out his similarities to other members of the animal kingdom. Since it is also implausible to suppose that man's essential characteristics are that he lacks feathers and walks on two legs, it fails as a realistic classification; Aristotle's 'political animal' still seems as good an essential definition of man as any. In his biological works, Aristotle made some of his major classes depend upon one character – such as the presence or absence of red blood, producing a division very close to ours between vertebrates and invertebrates – but in general used multiple criteria.

This enabled him to deal with some anomalous creatures, such as the bats, most of whose characteristics are mammalian and which therefore go in that class; and the dolphins, to which the same applies notwithstanding their resemblance to fish. Aristotle gave more weight to some characters than to others, paying especial attention to reproduction; but when he found and dissected a dogfish which, like the dolphin, seemed to bear its young alive, he still recognised that the bulk of its characters put it clearly with the fish. For the most part, then, Aristotle's system was based not on exclusive characters but upon family resemblances.

Family resemblances became interesting to philosphers in the twentieth century because Wittgenstein drew attention to them. Games like football, solitaire, snakes and ladders and chess seem to have little or nothing in common that entitles us to group them together; but the word 'game' is used to cover a family of activities each of which display some characters from a long, and perhaps indefinite, list. If more philosophers had studied the life sciences, and looked at Aristotle's biological works, instead of confining their interest in science to pure and applied mathematics, with perhaps a dash of experimental physics, Wittgenstein's remarks would not have seemed such an amazing bolt from the blue, but simply a bold extension of an established way of proceeding into new territory. When we go to a wedding and see the bridegroom in the midst of his sisters, and his cousins, and his aunts, we cannot fail to perceive family resemblances: one person has ears like the member of the family we know, another's hands look familiar, another stands in the same way, and so on. One could draw up a list of family characteristics; but nobody would possess them all, and there might even be members of the family with several characteristic features but none in common.

This last suggestion takes the analogy of family resemblances further than those who had used it in natural history or in chemistry. The idea of families used in these sciences was, down to the mid nineteenth century when Darwin took it seriously, a metaphor – like the term 'affinity' in chemistry, or 'charm' in nuclear physics – which could be taken too far, as it would be if one were to group together entities having nothing in common. Scientific families are composed of members each of which has to display a majority of those qualities which characterise the family. Underlying this procedure is the unproveable metaphysical hope that nature really is ordered into unambiguous families. This hope that there is an intelligible order in nature is like, and indeed is probably a version of, other general metaphysical propositions that make any kind of knowledge worth pursuing: that every event must have a cause, and that nature is uniform.

The justification of all such propositions is a part of the problem of induction; induction being the process by which we go from particular instances to general rules, that is whenever we generalise or put things into classes. The problem was put nicely by Bertrand Russell in a story about a chicken, which a man came and fed every morning; after many mornings, the chicken concluded that it was a law of nature that the man fed it every day, but the next day he came and wrung its neck. How do we know that we do not live in an unpredictable world like that of the chicken? Unlike the chicken we do know with Lord Keynes that 'in the long run we'll all be dead', but that's not much consolation. We cannot appeal to any principle of the uniformity of nature to reassure us, because it is that very principle that we are seeking to justify; and we have to agree that it is an act of faith to assume that past regularities will continue.

It is up to philosophers to have neurotic worries about whether the apparent order of the world might prove illusory. In ordinary life, and in science, we take it for granted that there is some order and that we can and have found out something about it; our concern is that some causal relations and some classes are only apparent. There is a famous relationship, followed over a large number of years, between the death rate in a part of India and the membership of an American trade union, oscillations in the one being paralleled in the other just as changes in the volume of a gas accompany changes in its pressure. And yet the first relation must be accidental, while the second is believed to be somehow real or even necessary. There is no way in which we can be completely certain that our best-established generalisations are not just accidental; what we do is to put more reliance on those which have been observed a great number of times, and especially on those which are connected to a whole series of others in the structure of a science, where they must to some extent stand or fall together.

What applies to generalisations applies also to classifications. Here the distinction is not between real and accidental causal connections, but between natural and artificial systems. Because the word 'system' had come to acquire overtones of dogmatism and impatience with mere evidence by the nineteenth century, opponents of artificial systems preferred to describe themselves as devotees of the natural *method*; but we may continue to use the useful word 'system' for both. Artificial systems have an *ad hoc* character, but they may still be useful and indeed may be very effective – in this perhaps they differ from accidental generalisations, and resemble some 'purely empirical' equations in physical chemistry. Just as empiricists would say that we can never be certain that any generalisation is a law of nature, and that we should recognise the inevitably provisional character of all scientific explanation, so nominalists would urge us to stop at a convenient and effective system of classification without trying to answer the unanswerable question of whether it is the real order of nature.

There is no easy way of dealing with such sceptics, who can after all point to Newton's theory of light-particles or to the Great Chain of Being of the eighteenth century as two examples among many of apparently well-based theories or sytems that have been abandoned. The same may be expected to happen with current theories of gases or classifications of spiders. If earlier theories had been confessedly provisional and adopted only for convenience or economy of thought, then the trauma of coming to terms with scientific change would have been less. What can perhaps be said is that this is an overcautious way of going about things, and that faint heart never won fair lady. If there had not been bold spirits searching out Nature's secrets and putting her to the question, then we might have had a certain amount of technology, horticulture, and medicine: but we would not have had, for example, Darwin's theory of development which depended upon both the work towards a natural system of classification in the early years of the nineteenth century, and the simultaneous attempt to explain how the processes of geology have formed the Earth we live upon and to classify the strata in terms of their fossils rather than their composition. The really interesting and exciting parts of science are those where a real attempt is made to find out the order of the world.

One of the leaders of the Cambridge school of physicists in the nineteenth century, Sir George Stokes, remarked that theory should not merely link up observations, but should serve to indicate the real processes by which nature brings about effects. He and his associates were thus concerned with the whole order of things, and it was no accident that one of them, Lord Kelvin, should challenge Darwin on the grounds that the second law of thermodynamics (of which he was a discoverer) did not allow time enough for the whole evolutionary process to have been brought about by natural selection. If truth is the objective, then theories must be extendable far beyond the range of phenomena (the performance of steam-engines, in Kelvin's case) for which they were originally devised. Truth is indivisible. In contrast to this, chemists in the nineteenth century who for the most part used but did not believe in an atomic theory, were not interested in theories of heat and of gases formulated by contemporary physicists even though to later generations these seemed excellent evidence for atoms. If one simply wants something that works in one's own field of interest, why should one pay much attention to what is going on over the hedge? It is a pity that nowadays academic frontiers – between chemistry and physics, and also between physics and ethics, for example – seem to be efficiently policed, so that more of us are like the nineteenth-century chemists than are like the Cambridge physicists.

In just the same way, a natural classification must be in some sense absolute. It must show where things really belong. Just as there can be radical but unalarming disagreement between theories of heat used by chemists and physicists if both sides see them as mere tools, so there can be any number of artificial classifications that cut across each other. The way to the natural system is, by contrast, strait and narrow; any real exception is fatal to it, or at least involves serious modification, while an artificial system can merely have its limits drawn a little tighter. In the search for the natural system the stakes are high; one may go quite wrong and find that one has wasted most of one's working life. This happened, for example, to those chemists in the nineteenth century who determined atomic weights with ever higher accuracy, believing that they were finding fundamental constants of nature that would enable the elements to be classified unambiguously. In twentieth-century atomic theory it is the charge on the nucleus that is this fundamental quantity, and the atomic weight merely represents the ratio of the isotopes in the sample, which will usually be constant but, with lead, for example, is variable. Very precise atomic weights were not really worth spending a lifetime on.

The artificial nature of some systems is clear enough. We may divide fish into those that are good to eat and those that are not, and we may divide foods in general into those that are fattening and those that are not. Here we are clearly imposing order on the world rather than discovering some fundamental way in which it is ordered; and our purposes are clear – artificial classifications are justified by their practical end. There is no reason why the system should be two-valued; there were some extensive official reports published in the USA in the nineteenth century, still of value to entomonogists, which set out to describe the 'noxious, beneficial, and other' insects found there. A surprising amount of natural history has been done as a contribution to hunting, horticulture, and pest-control; but there is usually a gulf to be crossed between this kind of practical knowledge and the more speculative part of the science concerned with natural classification.

An important aspect of classification is naming. Once we can give names to things we have come some way towards ordering them; and the names may even indicate the place things have in the order. Thus a name ending in 'um', such as potassium, indicates a metal and this is why Davy chose it. Names of chemical substances in the eighteenth century generally lacked this feature; thus the metals that have been long known include silver, copper, gold, nickel, and zinc and there is no clue in these names that all the substances named belong to one class. It was Lavoisier and his associates in the 1780s who set out to reform the nomenclature of chemistry in line with philosophical principles. Lavoisier wanted to cut the science adrift from speculations about the nature of matter, which must be unproveable, and turn it into an organised and firmly-based descriptive science; he carefully defined the term 'element' and drew up a table in which the elements were placed in various classes. The names given to the newly-discovered ones indicated their nature: thus 'hydrogen' meant 'generator of water', and 'oxygen' meant 'generator of acids'.

Oxygen was in a sense the centrepiece of Lavoisier's chemistry, for it reacted with almost everything and he had been the first to interpret its reactions. But although one can see some connection between oxygen and acidity – where metals form several oxides, those with least oxygen are basic and those with most are acidic – it was clear within twenty-five years that there were many acids that did not even contain their so-called generator. Clearly, despite his intention to avoid theory-loading, Lavoisier had passed on to his contemporaries and successors a nomenclature that enshrined a mistake.

The greatest exception was the acid made from common salt, in which Humphry Davy – the great chemical genius of the early nineteenth century – found it impossible to detect any oxygen. Nor could he find any in the greenish gas that could be prepared from this acid or its salts, and which had been called 'oxy-muriatic acid gas'; since it was not acidic, and did not contain oxygen, Davy renamed it 'chlorine' and placed it among the elements. This name referred only to its colour (it had been used by Linnaeus for the greenfinch), and could therefore be retained, Davy thought, through whatever changes in chemical theory the future might bring; and so it has proved. Davy had previously been among those who had rejected the French term 'azote', meaning destroyer of life, in favour of the theory-free word 'nitrogen' (from its relation to nitre) because his work had indicated that nitrogen was not poisonous (like chlorine) but merely suffocating. Lavoisier had played for the high stakes of a nomenclature which would reveal the essential nature of the things named, and had lost. Davy's more cautious attempts at family groupings, and at theory-free names, has stood up better; though his proposal for naming the compounds of chlorine in a kind of shorthand did involve some theory and a good deal of innovation and was never widely adopted, even by its author.

Davy did use names to display analogies, as when he coined 'iodine' by analogy with 'chlorine' to bring out the similarities between these substances; his French contemporaries and rivals using 'chlore' and 'iode' made this point less forcefully. It is not only in chemistry that there have been difficulties about the theory-loading of terms; in natural history, throughout the eighteenth century in Britain and in France, our own class of mammals was generally described as the quadrupeds, and indeed this was the term used by Cuvier in his great works on living and fossil mammals published in republican and Napoleonic France. Most mammals are indeed quadrupeds, but man is not and nor are seals, dolphins or whales; while on the other hand lizards and frogs are quadrupeds but not mammals. By a rather different process, the terms 'reptile', which in the eighteenth century meant anything creepy-crawly (and therefore made an excellent term of abuse), was narrowed down so as to exclude invertebrate creatures like centipedes by the end of the century, and 'amphibia' was similarly refined by the middle of the nineteenth. 'Insect' in the early eighteenth century had been synonymous with 'reptile', as in Lawson's *History of Carolina*, 1709; and it still was in Victorian railway taxonomy, for Frank Buckland the naturalist found that he must pay for a monkey which counted as a 'dog', but not for a tortoise which was an 'insect'.

We have been preoccupied with scientific naming, but in fact all naming involves classification. During the Rennaissance, Adam in the Garden of Eden before his vision was clouded by sin was supposed to have been able to perceive the essences of things and thus to name in accordance with nature. His original language had been lost at the time of the Tower of Babel, but it could be presumed that Hebrew was closer to it than were the tongues of northern Europe. Whereas in English or German, then, names were mere noises to distinguish objects, Hebrew names somehow came close to disclosing the essences of things, giving a magician power over them, for example. Magic, rather than experimental science, was the great preoccupation of those interested in understanding and getting power over nature in the early seventeenth century; and this interest only waned slowly as the rationalism of Galileo and Descartes spread, as the example of Newton's alchemical studies may remind us. But magic worked only moderately well – perhaps as well as psychiatry or meteorology in our day, some magicians making good counsellors and forecasters – and could not be fitted in with the persuasive mechanical and rational worldview that we think of as a characteristic of the eighteenth century. With belief in magic departed the belief that one language might be better than another at penetrating to the essence of things; the way to classify naturally was not to look backwards to some earlier and supposedly more enlightened period but to go forwards in confidence that the world contained no mysteries that the human mind could not master.

It was generally supposed in Britain that the categories according to which everything was classified were to be discovered; whereas in France there was more readiness, in the tradition of Descartes, to believe them innate in the human mind. At the end of the eighteenth century, the great philosopher Kant urged that we impose categories on nature as the way of making the world comprehensible: at different times and places people have imposed different categories; but unless we get appropriate categories we shall not be able to order the world. Our experience will indicate the appropriateness of the categories we use, which can be revised within limits as we go along: the categories determine what we experience, but are in their turn modified by it. Kant came out as an opponent of the mechanical philosphers who sought to reduce everything to machinery, looking on the world as a kind of overgrown clock; for Kant different kinds of explanation were appropriate in different fields, and he saw the sciences as irreducible to one another and arranged in a kind of hierarchy.

Any classification is thus composed of elements that we impose and elements that we discover; it is the result of mind working on matter, and the imposed categories can with experience be refined but never eliminated. Like everything else interesting, classifying involves judgement, and in matters of judgement even men of goodwill can disagree. The way to a truly natural system involves both the collection of relevant material (relevance being determined by one's categories) and also the refining of the principles upon which the classification is based. The past is full not only of acknowledgedly artificial systems that had their day of usefulness and have now become obsolete, but also of systems which set out to be natural but turned out to be artificial, grouping unlike and separating like.

We shall come upon many of these when we look in due course at the various classifications that have been used in the life sciences and elsewhere. In science the general difficulty remains, that there is no clear way in which one can be sure that any given system is not imposed and artificial, based on a few superficial or misleading characteristics, rather than really discovered and natural. Although classification in the sciences is carried beyond what is done in ordinary life, and the distinctions become more recondite as the problems become more technical, it is in our ordinary life as political animals – that is, ones that live in communities – that we continually classify.

Human societies change over time, and it is no longer true in Britain as it probably was a hundred years age, that 'every child that's born alive, is either a little Liberal or else a little Conservative'. That classification no longer works; though it might perhaps do so if the categories were 'liberal' and 'conservative'. We classify people as students, as pensioners, as capitalists; as élites, as the working class, as the proletariat; by sex, by age, and by colour. Making these classifications – which we may feel reluctant to do but neverthless cannot stop doing – involves social and political judgements. How one places oneself in society, and how one sees the forces that make that society work, will be associated with one's recognition of some of these categories as natural and helpful in ordering the world, and one's rejection of others as artificial.

Classifications that seem to work in society seem to be paralleled by those that work in science at the same time. Thus in the eighteenth century, the most favoured system of classifying living organisms was the Great Chain of Being, where all creatures were placed like links in a chain, stretching from minerals, through vegetables and animals, up to man, and perhaps on through the angels. In such a scheme, every creature had its place, and any attempt by one species to move up a link could only produce disruption and chaos. This is a scheme which resembles an ideal *ancien regime*, with its various strata, from the landless peasant through the gentry and nobility to the king; and one would expect that it would recommend itself to those who perceived society as hierarchical and static. By way of contrast, the middle of the nineteenth century was a confused and unstable period of social struggles. The great naturalist and geographer Alexander von Humboldt, who was born in 1769, had been in Paris during the Revolutionary and Napoleonic years, lived on to see that progress can be stopped with the abortive revolutions of 1848, and died in the same year, 1859, that the *Origin of Species* was published, still regarding himself as 'a man of 1789'. To those coming to maturity and studying science in the late eighteenth and early nineteenth centuries, society bore an unstable aspect; institutions had their lives, and struggles were the rule. The struggle for existence – a term coined by the proto-sociologist Herbert Spencer – seemed a more obvious pattern for nature then anything associated with unchanging hierarchy. Darwin's theory was therefore in accord with the spirit of the age; the more so as Darwin had to allow that some kind of progress had clearly happened in evolution, even though there seemed to be no necessity for progress in the operation of natural selection. Both those confident of progress and those uncertain whether it could continue could find support in biology.

Darwin's theory seemed very compatible with *laissez-faire* economics, and 'social Darwinism' was a feature of right-wing politics in the late nineteenth century; but prominent Darwinians were to be found politically on the left, the great example being A.R.Wallace who wrote in favour of land-nationalisation and socialism. Indeed, in formal and theoretical science, which must be the product of a complex and open society, it is hard to get beyond a very general sketch of how certain ways of thinking about nature might reflect similar ways of thinking about human relations. Detailed studies always seem to indicate that nobody ever was a typical mid-Victorian (or a typical anything else): and to link attitudes that we might think of as characteristic of an era with the science of the same period is very difficult if we try to get beyond generalities. It might be possible to do it for simpler societies, making simpler classifications; and indeed in 1903 the sociologists Durkheim and Mauss wrote an essay on primitive classificication in which they argued that classifications must mirror social relations.

On this view, there is not just a parallel or an association, which seems usually the most one can see with science in complex societies, but a causal relationship; so that a society having a certain structure will classify in a certain way. Their essay was a piece of armchair anthropology, based largely on field reports on Australian aborigines and American Indians, sometimes mistranslated and misunderstood; and the conclusion of their recent editor and translator is that in none of the cases cited 'is there any compulsion to believe that the society is the cause or even the model of the classification.' The causal relationship will not do in the simple way proposed; but the idea of looking at features of a society to account for features of a classification has turned out to be very interesting – even though some rather different societies have come up with rather similar classifications.

One problem is that cultures and social systems are not as tidy as we might hope classifications to be; we know this for ourselves, but it seems to be true of other societies also. Thus F.E.Williams, who was a government anthropologist in Papua New Guinea, proposed in place of the view that a culture was a coherent system like a machine or organism in which the removal of any part would bring about complete collapse, the model of a culture as something like a rubbish heap. All the parts are loosely held together, but not in any carefully articulated structure; and one can pull out the equivalent of an old tin or two without the whole heap tumbling down. Such a loose texture is a feature of relatively informal classifications, too, and one should not expect, therefore, to find a close fit between them and social systems as one might between carefully-defined variables in a scientific experiment.

There is also the difficulty, as Mary Douglas pointed out in her fascinating book, *Purity and Danger*, that the observer must interpret classifications, rituals and so on. Her own interpretation of the 'abominations of Leviticus', the dietary rules of the Old Testament, themselves indicate this, because after all these are passages well-known in Christendom which have baffled the ingenuity of commentators trying to find in them some evidence of the working of God in the world. She sees the Israelites, as always in their history, a hard-pressed minority, the threatened boundaries of their body politic being 'mirrored in their care for the integrity, unity and purity of the physical body.' The priests, and the sacrifices, must be especially pure and perfect; but the diet of the whole people must also have this integrity, and they must avoid anomalous and unclassifiable creatures. The pig is one of these, for while it has cloven hoofs it does not chew the cud as such an animal should do; it is therefore not a proper sort of creature, and is to be left alone like the first armadillos in the Just-so Story. The same sort of anomalousness is a feature of all the 'unclean' beasts, and when the rules are seen as a part of a classification system, with a close connection though not a causal one to the social system, then they have a logic which is otherwise not apparent, whether they are believed to be rules of hygiene or tests of obedience.

The thread running through Mary Douglas' book is that the borderline is always an alarming and dangerous place; that when a society has classified people and things, then what fails to fit comfortably into the grid may seem frightening or polluting. Ideas of what is unclear are one side of a coin, the other side of which is notions of order and pattern, and classification is fundamental to what might otherwise seem very curious doctrines and practices. These borderlines run not only between animals with cloven hoofs and those without, but also between adults and children and between the living and the dead; in our own society these borders are daunting places with the problems of adolescence at one, and those of euthanasia and support systems, and of abortion, at the other. Quarrels in both cases arise when there is a difference of opinion over whether somebody is really grown up, or whether a person or foetus is really alive.

Within the sciences too, borderlines are hazardous places. The history of semi-conductors, to which are now the basis of the electronics industry, owes its surprisngly late start to their marginal status in the late nineteenth and early twentieth century spectrum of the sciences. The study of gases had been respectable since the work of Dalton and Gay-Lussac soon after 1800 (if not that of Boyle over a century before that) as a quantifiable field where clear experiments and simple laws might be expected, and where the physicist had a role distinct from that of the chemist. All gases behave alike when subjected to changes of temperature and pressure; they seemed therefore simpler than liquids or solids; or, in taxonomic terms, they formed a natural group. Liquids and solids behave in all sorts of different ways under different physical conditions, and there were no tidy quantitative laws to be found, or so it seemed, by nicely repeatable experiments.

By 1859 Maxwell and Clausius had made the kinetic or dynamical theory of gases, according to which they are composed of particles colliding elastically and moving on average faster as the temperature rises, as well-established as most parts of physics. It was not surprising that the electron should be identified in J.J.Thomson's work of 1897 in a tube containing gas at low pressure; and that early radio and television sets should have had valves which, like Thomson's apparatus, consisted of almost-evacuated tubes containing various electric terminals. Gases were understood, and the passage of electricity through them was reasonably predictable. Research students would be well-advised to follow this line of enquiry, a central part of the science that could hardly fail to generate interesting problems which there were techniques for solving.

Solids were another story. Chemists had indeed drawn up tables of elements and compounds, and crystallographers had worked on the forms of crystals; but around 1900 the solid state was not understood as the gaseous was, and in particular there was no real glimmering of a theory of insulation and electric conduction. Some mysterious substances had indeed been investigated; thus selenium had been tried out as a very high resistance for telegraphy, and it had been noticed that the resistance varied when the substance was exposed to light. This property of selenium was to be used in the twentieth century in exposure-meters and in the xerox machine; but in the nineteenth century, although it at first attracted the attention of some men of eminence, the results proved hard to interpret and control, the investigation did not pay off in the appearance of any law or theory, and the field was left to a few inventors and men of science out of the mainstream of physics.

The very class of 'semi-conductors' was not defined until well into the twentieth century; and the problem was that this class cut across well-established classes. The first of these was the metal/non-metal class: silicon we have already met, apparently straddling this boundary, and selenium is also a substance on the margin between metals and non-metals. The next boundary was between elements and compounds: selenium is an element but

31

semi-conducting properties are also shown by some sulphides and selenides. Nineteenth-century standards of chemical purity were not as high as those required for producing materials for semi-conductors today, and the behaviour of different samples varied widely. Both semi-conductors themselves (which of course again meant neither conductors nor insulators), and the whole science of the solid state, seemed marginal and treacherous until, in the late 1920s, quantum theory, devised to deal with emission and absorption of light, proved able to account for the passage of electricity through solids, and ultimately to make the process look simpler than that through gases. From a margin, these phenomena moved into the middle regions of physics, and became safe.

One of those who, in the second half of the nineteenth century, did some of the original work on the passage of electricity through gases was William Crookes; some of his experiments on cathode rays are classics. In the scientific community he was, as J.J. Thomson later described him, a marginal person: he had not had a scientific education (but nor had several of his contemporaries in science), he did not hold an academic post, he had rather flamboyant moustaches, and he had owned a gold mine; he supported himself by scientific journalism and consultancy, and he entered a scientific field like an explorer recording everything he noticed and not like a trained physicist with a theory to test. As Crookes himself said, his interest was in the borderland between the material and the immaterial; this led him to investigate the cathode rays, but it also led him (like many late Victorians of eminent intellect) into psychical research. This work on the boundary of science and religion proved to be rather like the investigation of selenium: one could never get fully consistent results, or assign real theoretical significance to what was observed, partly because it cut across accepted categories of causation. There was in addition the problem of fraud; but the work of Crookes and of the early members of the Society for Psychical Research is exceptionally fascinating because one sees hard-headed experimenters trying to come to terms with data that do not lend themseves to the methods of physics or chemistry.

It would be hard to relate the cultivation of a physics of gases rather than solids to the social system of the nineteenth century; but for psychical research it seems not too difficult, for the churches were in retreat before scientism; and it seemed as though to make life meaningful, some experimental proof of the survival of death was needed which could replace the message of Easter. One does not need much elaborate justification for interest in a science which touched upon such fundamental uncertainties of mankind; but, concerned with a borderline, and unable to find an appropriate framework of categories, psychical research gradually lost its following of eminent scientists, philosophers, and intellectuals, remaining, unlike solid-state physics, a marginal activity. It does no good to anybody's reputation nowadays to have been involved in it.

It is not usually easy to examine two systems of classification which order the same things (more or less) and which are the products of two different cultures. But this can now easily be done for the birds of paradise and bower birds. In 1969 Thomas Gilliard's monograph on these groups was published, followed in 1977 by a sumptuous volume with text by J.M.Forshaw and plates by William Cooper, and also by *Birds of my Kalam Country* by Ian Saem Majnep and R.Bulmer, with handsome plates by C.Healey. The first of these is particularly concerned with the taxonomy, the behaviour, and the evolution of these fascinating groups of birds; one of which has developed extraordinary plumes and elaborate dances to show them off, while the other builds and decorates – sometimes even using a stick as a paint brush – elaborate structures called bowers where the male entertains the female, but which is not a nest where eggs are laid. Forshaw and Cooper set out to complement Gillard's book with superb illustrations and a text brought up to date and including some new interpretations. The third book is written by a native of New Guinea with an anthropologist.

Gilliard was interested in the birds of paradise because of their specialisation, distinct species being found only a few miles apart; because they are polygamous, unlike almost all other birds which pair; and because hybrids are found not only between different species but even between different genera. These facts seemed to indicate that very rapid evolution was going on in the group. In the bower birds he discerned a further step: sexual selection had produced the plumes of the birds of paradise, for they attract the females and in a polygamous group the more females one can attract the more descendants having one's characters one will leave. In bower birds the males are much plainer, and here the sexual characters seem to have been transferred to the bower and its contents. Thus the male woos the female by displaying his collections and works of art; and as Gilliard remarks, this seems to indicate some evolutionary leap, similar to that which must have happened in human evolution, towards the use of tools and things made with them. His work on these groups was therefore guided by Darwinian theory – Wallace having actually been a pioneer in the study of these birds – and led, via some rather technical studies like those of the birds' skulls, which put them in the starling family, to greater understanding of how evolution works and even perhaps how a stage in human evolution may have taken place. It would be hard to say that his classification mirrored a social system, but it does fit in very well with a world-view that has guided biologists for a century; and Forshaw is in the same tradition.

To the Kalam of New Guinea, birds of paradise and bower birds are only part of the local fauna, even though they are important for their plumes worn for festivals. Saem recounts stories about the various species of birds, which include some natural history and some folklore. This would be like stories of Jenny Wren and Cock Robin in Britain; and where Saem's birds have symbolic importance, this would be like our calling the eagle the King of Birds. Davy, when President of the Royal Society, wrote a poem about an eagle teaching its young to climb upwards and gaze fearlessly at the sun, using it as a symbol of the role he hoped to play in encouraging the young to take up science; we still use animals symbolically, classifying them as kingly or cunning for example, as well as fitting them into Darwinian categories.

The Kalam name the birds too, using some general characters and some specific ones; and Bulmer has tried to tabulate the agreements between their system and ours. He found that unlike us they put bats in the same group with flying birds, and put the cassowaries into a distinct group. The great majority of their names did correspond to species or higher groups in our system; where they did not have a name for some kind of bird – probably a kind they rarely saw – they usually put it into an appropriate group. Out of 157 Kalam names, Bulmer found that ninety-six applied to species, while nineteen applied to groups of species (thirteen of these being valid taxa in our biology, and six being groups only superficially similar), twenty-nine corresponded to divisions within species (often to the different sexes), and eight to parts of more than one species (as the similar females of different birds of paradise), while only five remain unidentifiable. This seemed to demonstrate, as Bulmer says, an astonishing degree of similarity between the classifications of two extremely different societies with different interests in birds, and must encourage those who believe that an order is to be found in nature and not just imposed there by different societies. The higher groupings do not always reflect ours: some birds of paradise are in the group 'in which women show themselves', others in those 'that feed on certain fruits', and then the yellow and red birds of paradise as being different from all others are in a group of their own. We find in fact a symbolic, an ecological, and a systematic classification mixed together.

All systems are likely in some sense to reflect their culture; with those of concern to all, the society will be in some degree mirrored in the classification, while more technical categories will demonstrate the preoccupations of scientists, But the Kalam and their bird classifications, and the steady development of biological classifications begun in the city-states of Ancient Greece and carried on through the Renaissance, the *anciens regimes*, and into the modern world, must give us the hope that we are in some sense finding the order rather than seeing ourselves or our society reflected there and nothing more. A natural system without any misfits or classifications that cut across each other may be a dream, but it need not be a chimera. Mary Douglas writes that 'now that we have recognised and assimilated our common descent with apes nothing can happen in the field of animal taxonomy to rouse our concern.' We have now to move back in history to a time when this was not true; and when the search for a natural and objective system always carried a threat (or offered possibly valuable gifts) to our culture.

2

An Objective System

———

ONE OF THE fundamental distinctions that were made, up to the Renaissance, was between the earthly and the heavenly. The birds of paradise were so called not only because of their striking beauty, but also because they were supposed to have no feet (these were removed in Indonesia in preparing the skins) and to spend their whole time in flight without ever touching the Earth. Saint Augustine in his *City of God* described the two cities, the earthly and the heavenly Jerusalem. In general, the heavenly came to stand for the ideal and unchanging while the earthly was the sphere of imperfection, change, and decay. We could get on Earth glimpses of Heaven through liturgy, through beautiful creatures, things, and works of art, through saintly men and women, and from contemplating the starry heavens. The creation gave abundant evidence of its creator; the earthly or profane was not utterly separated from the heavenly or sacred, and the study of nature could not be divorced from that of man and God.

Around the Earth, the centre but also the rubbish heap of the creation, revolved the heavenly bodies. There were two categories of these: the 'fixed stars' which all moved together around the Earth each twenty-four hours, and the 'planets' or wandering stars which went their own ways. These were the Moon, Mercury, Venus, the Sun, Mars, Jupiter and Saturn, making the number seven which stood for completeness. To us, they are a curious class, varying as they do in brightness and in importance to man; but the world was seen as animated by spiritual beings, and astrology was believed in by the educated down to the end of the seventeenth century, and therefore the influence of Venus, Saturn, and the rest was as important in its way as that of the Sun. All the planets moved through the constellations called the Signs of the Zodiac along a line called the Ecliptic inclined at about twenty-three degrees to the Equator; we are familiar with this with the Sun, for it comes up to 23°N in June, passes over the Equator in

36

September, gets to 23°S in December, back to the Equator in March, and so on. Planets did wander, and they moved irregularly (the Sun for example spends longer in the northern than the southern hemisphere) but they could not go just anywhere, being confined to a band of the sky; and this indicated that they properly belonged in one group.

Babylonian astronomers had tabulated these movements and had realised that the heavens were a kind of gigantic clock in which everything happened in cycles: once the full cycle had been followed, one could know what would happen next, and indeed they developed mathematical tables for making predictions. Ancient Greek astronomers developed explanatory models to account for what was happening, for them eclipses were not just curious phenomena, the possibility of which could be predicted from tables, but resulted from actual bodies getting in a line and obscuring one another. For Plato the divine nature of the heavenly bodies was shown by their regular behaviour, and he rejoiced when Eudoxos demonstrated that the apparently irregular motions of the planets could be composed of a number of regular motions in circles. Then as now, one could debate whether it was more divine to be completely predictable and dependable, the eternal ruler of the ceaseless round, or to be the unexpected lord of the dance: for men of science, as for Plato, God has generally had the former attributes, reigning over the best of all possible worlds, while any marginal, disturbing and innovating power would be looked at askance, like Maxwell's Demon that could break the Second Law of Thermodynamics.

Aristotle's physics enshrined various classifications which were taken for granted in the west down to the seventeenth century. Celestial bodies were made of different stuff from earthly ones, a kind of quintessence with different properties from the earth, air fire, and water which made up all that we are familiar with. Ordinary matter had its natural motion; up (in the case of fire), or down (for everything else) towards the centre of the Earth and the world; but the natural motion of quintessence is circular, being the next best thing to rest which is the attribute of the Unmoved Mover, God. Planetary motion were composed not of simple circular motion, but of combinations of circles; Eudoxos' system was replaced by Ptolemy's, which involved such complicated mathematics that King Alfonso X of Castile in the thirteenth century is supposed to have said (when supervising the preparation of astronomical tables) that if God had consulted him about the creation of the heavens he would have given Him better advice.

Copernicus in the sixteenth century showed that the mathematics were no more complicated if the Sun were put at, or rather near, the centres of the planetary orbits, and the Earth put into motion around it. In some ways such a system was simpler, for it placed the Sun in a separate category from other planets and at the centre where the source of light and heat and thus of life should perhaps be, and it accounted for the apparently backwards motions of other planets as simply the effect of the Earth overtaking them or being overtaken by them in its orbit. On the other hand, it upset all sorts of accepted laws of physics such as that of natural motion, and raised the problem of how the Earth stays in its orbit and how we stay on a swiftly moving and rotating Earth. It even seemed to be falsified because one would expect apparent motions among the fixed stars as the Earth swung round its enormous orbit, but none could be detected: to good, sound men, Copernicus' answer – that the stars must be very much further away than anybody would have thought – sounded rather lame. The pre-Copernican world was already tens of millions of miles across, and a jump to thousands of millions of miles seemed uncalled for.

The Copernican system therefore seemed a rather wild hypothesis or just a mathematical model, and as such it did not threaten anybody's world-view. Philosophers had already developed the doctrine of the double truth, whereby real truth was to be found in the Bible and the traditions of the Church, while in the sciences and other practical activities one could be satisfied with a kind of pragmatic truth – if a theory worked and was consistent with itself, then there was a sense in which one could say that it was true. These merely coherent and pragmatic propositions or classifications should not be put on a level with revelation; but in their own sphere, and especially in books written or lectures delivered in Latin for experts, there was no problem even if they seemed to clash with a straightforward interpretation of Scripture. Copernicus' work had had the approval of Church dignitories and was dedicated to the Pope; it had, moreover, an unauthorised and unsigned preface by the publisher explaining that it was no more than a hypothesis. Copernicus had also put on the title page a Greek phrase said to have been over the door of Plato's Academy, meaning something like 'non-mathematicians keep out'; and so the categories inherited from Aristotle were not seriously challenged.

It was Galileo who played the role of Copernicus' bulldog in placing the Earth firmly among the planets. In 1610 he published, in Latin, his little book announcing his discoveries with the telescope, *Siderius Nuncius* – the Starry Message or Messenger. Seen through the telescope the Moon was clearly a rocky and mountainous mass, not made of some quite different kind of matter from the Earth, and not a smooth sphere. The Moon manifestly circled the Earth, as the Sun and the other planets seemed to do, and it had seemed implausible that, as Copernicus had urged, there were two centres of motion in the world: the Sun for the Earth and the other five planets, and the Earth for the Moon. Galileo with his telescope saw the moons of Jupiter, and when others doubted he offered a prize to anybody who could make a telescope that would make planets seem to go around Jupiter but not around anything else that he pointed it at. The Earth was therefore not unique in having a satellite; rather this was an analogy between the Earth and another of the planets. Whether one liked it or not, there were several centres of motion.

It was a feature of the planets that they shone – that indeed was all that was known about them, apart from their paths across the sky. The Earth seemed quite different; we know that it does not shine, for if it did we would not need lights at night. But Galileo was able to interpret a known phenomenon so as to indicate that the Earth does shine like the Moon; that it scatters some of the sunlight which falls on it back into space. When the Moon only shows as a tiny crescent, the rest of its surface is on a clear night seen to be dimly lit; the phenomenon is sometimes called 'the new moon with the old Moon in her arms'. At new Moon, the Moon is approximately between the Earth and the Sun, and Galileo argued that this dim light is Earthlight scattered back from the Earth's surface and illuminating the Moon just as Moonlight from the bright crescent is giving us light.

The Earth and the Moon were therefore similar, in that they were not self-luminous, but both shone when the Sun shone on them. The next problem was to carry the analogy to the planets, which Galileo did with observations on Venus. He found that she showed phases like the Moon. As she circled the Sun, when furthest from the Earth she appeared a complete disc, and when nearest to us a crescent. This could only be accounted for if she (and by analogy the other planets) did circle the Sun rather than the Earth, and if she was not self-luminous. Seen through the tele-scope planets and stars also fell into distinct groups; the fixed stars were not magnified at all, indicating that their distance must be enormous (as Copernicus had argued), while planets showed as a disc (except for Venus, which showed phases). Stars and the Sun, therefore, for Galileo formed one class of bodies, widely-spaced and self-luminous; while planets and their satellites formed another. In superficial ways stars and planets seemed to resemble each other – both were bright spots in the night sky – rather as birds and bats, or fish and whales do; but when fundamental properties like optical behaviour were taken into account then planets, including the Earth, formed a natural group.

We sometimes think of Galileo as the great pioneer of an exper-imental and mathematical physics, which indeed he was; but his achievement was as much taxonomic as explanatory. He had no satisfactory theory of gravity, and had to suppose that the circular motion of planets in their orbits (for he never believed in Kepler's new-fangled ellipses) was 'natural', even though he had come some way towards replacing Aristotle's laws of motion. In his book of 1632 on the *Two Great Systems of the World* (the very title of which rouses echoes of taxonomy) he spends much time arguing for this reclassification of the Earth, presenting its analogies with the planets and going on to say that a body which agrees with them in so many characters must agree with them in others, such as going round the Sun in an orbit. He could not demonstrate this, and his argument which he intended to be a knock-down one from the existence of the tides to the rotation of the Earth was fallacious; he could not believe that the Moon could have any effect on the tides any more than planets could exert astrological influences; and, living in nontidal Venice, he knew little about tides anyway.

His book was in Tuscan, which was becoming the written Italian language, and was therefore open to people outside the academic community; this was indeed part of the offence that brought upon him the wrath of the Inquisition. Certainly his readers, who would not have been the sort of expert mathematicians for whom Coper-nicus had written, would have been more at home with his argu-ments for reclassifying the Earth and the Sun than with mathematical demonstrations however rigorous. Since the argu-ments were wittily expressed in a dialogue between interlocutors with real character, the book proved very effective. Like any work of theory or reclassification, the book could not demonstrate to the sceptic or conservative that the Earth must be a planet. Even Euclid's proofs only hold if one has accepted the axioms at the beginning of the book, and experimental or observational science must be a matter of probabilities, as, for example, Linnaeus and Darwin were to find in sciences apparently some way from astronomy.

To the argument of the inquisitors that the interpretation of Scripture and tradition could only be revised to accomodate new knowledge if it were proved to be true, there was really no answer. Critical reviewers of Darwin later made the same point, that they saw no reason why a satisfactory world view, unifying science with religion and morality, should be abandoned for mere probabilities. Then, T.H.Huxley, 'Darwin's bulldog', made the explicit point that, as a working hypothesis, Darwin's was excellent, because it suggested all sorts of questions while belief in special creations merely induced astonishment; for Huxley suggestiveness and openness was a positive advantage. For Galileo and his contemporaries, things looked different because they were trying to get away from a vision of science as likely stories and were in pursuit of certainty; and it was frustrating that the best Galileo could achieve was plausibility, explaining why if the Earth was moving we did not fall off but not proving that it was in fact moving.

If Galileo had been prepared to admit that one classification was as likely as another, then he and the inquisitors could have parted on good terms; but he believed that his was natural, and that the earthly/heavenly distinction had been shown in astronomy to be artifical. The Copernican Revolution in which Galileo played certainly the most public role has been taken as a triumph of objectivity. It seems obvious that the Earth is still, and that the heavenly bodies go around it, and that stars and planets are the same sort of things; but a careful and detatched investigation, by one prepared, as it were, to stand aside and not put himself at the centre of things, showed that the analogies were really very different. That we can admit, while noting that it would have been possible to go on with the old system; there would have been an increasing need for *ad hoc* devices and extra circles, and while the Aristotelian scheme could no doubt have accomodated a new planet like Neptune it could never (as Newton's did) have predicted it.

It has also been supposed that the Copernican Revolution in some sense dethroned man; but that seems much more doubtful. It was not until well into the nineteenth century that schemes of classification in zoology did not always focus on man as the summit of creation, other creatures being judged more or less perfect according to their degree of resemblance to him. In the pre-Copernican scheme, man inhabited a unique globe of matter, the Earth; there was nothing like it, and nothing like him, in the whole cosmos, and the drama of his salvation was really the purpose of the whole creation. On the other hand, the Earth was the sphere of change and decay, the dump as well the centre of the world, and to dwell on it was not a sign of dignity; the heavens were the abode of celestial beings superior to man, and in the highest heavens (furthest from Earth) was God. Man's place was therefore equivocal, as befitted a being coming between the animals and the angels.

If the Earth was one planet among others, situated in a vast and perhaps infinite universe, then things looked rather different. The size of the universe might either be taken as by Pascal as frightening, or rejoiced in because an infinite God would hardly create a less-than-infinite world; but even then there was the problem of why so much empty space had been left in it. The place of man suddenly looked more problematic, for if the Earth displayed so many analogies with other planets, why should they not also be inhabited? Galileo liked a fight but was prudent enough to keep out of that one; for men on other worlds were presumably not descended from Adam and Eve, and while Jesus' sacrifice had redeemed all mankind it might or might not be held to have saved the Martians and Venusians too. To postulate rational beings on other planets was to declare that man, like the Earth, was not unique, and that was courting trouble.

But in other countries there were those who were prepared to take this step. Fontenelle in France wrote, in 1686, what became the most famous work on the plurality of worlds, throwing the onus of proof upon opponents, and speculating about the amorous and negroid Venusians and the phlegmatic Jovians. He was a man of letters who became, on the strength largely of this book, the Permanent Secretary of the Academy of Sciences in Paris; he lived to be over one hundred, and was heard to remark, when stooping to pick up a lady's fan, that he wished he still had the elasticity that he had had at eighty. His older contemporary, the great astronomer Christiaan Huygens, wrote an essay published after his death as *Cosmotheoros* (1698) in which he took Galileo's analogies even further, arguing that the fixed stars must, as they are in the same class as the Sun, have planets too; which in their turn must also have rational inhabitants who must be very like us. The attempt to classify man with the hypothetical inhabitants of hypothetical planets seems decidely bizarre; but on the whole the doctrine of Fontenelle and Huygens seems to have been generally accepted through the eighteenth century, only to be assailed in the nineteenth when the place of man was again a live issue because of theories of evolution.

In 1854 Whewell published, anonymously, a *Dialogue on the Plurality of Worlds*, urging the hypothetical nature of all the arguments for inhabitants of other heavenly bodies. He argued that the Sun was probably not a typical star, and that others differed from it in colour, brightness, and in belonging to a system of two or more orbiting each other; so it could not be assumed that all stars had planets, let alone that such hypothetical satellites would be populated. The Sun, the Moon, and the solar planets all seemed to Whewell to be uninhabitable; and the soundest conclusion seemed to be that the Earth was unique, and therefore so was man. Any other conclusion involved extrapolating in a quite unjustifiable manner. Whewell was pleased that this conclusion was in line with the Church's teaching.

Whewell was quickly and acrimoniously answered by David Brewster, a Scot who was a great enemy of Cambridge men, in a work with the title *More Worlds than One: the creed of the philosopher and the hope of the Christian*. This work is rather breathtaking, because while arguments for the plurality of worlds have to rely on the uniformity of nature, Brewster denied uniformitarian theories in geology. He also believed that he had 'certain knowledge that the sun is not a red-hot globe, but that its nucleus is a solid opaque mass receiving very little heat and light from its luminous atmosphere', and there habitable, as William Herschel had suggested about fifty years before. It is odd that the ninth thousand of this book was being sold in 1862; this debate served to underline the role of analogy, and hence of classification, in the discussion of plurality of worlds. Nowadays many people are ready to accept the idea that there may be rational life elsewhere, though there is still no direct evidence for it at all.

Man's status in the universe may therefore have been called in question by the new astronomical taxonomy of Galileo, but as far as his terrestrial place was concerned it made no difference. He might have to share his attribute of reason with beings elsewhere (but after all, he already shared it with the angels, who were superior to him rather than equals), but he did not have to share it with he orang-outang, or admit any genetic connexion with any other kind of creatures. In Antiquity, Galen had often dissected apes when he could not get human corpses, assuming that they were put together in the same way; and Renaissance anatomists often used pigs similarly, without questioning man's uniqueness and distinctness from the animal kingdom. Renaissance philosophers concerned with man's place were not thinking about zoological taxonomy but about how a rational, but imperfect, being fitted into the order of things, and how he might by exercises or by magic gain spiritual power and true happiness.

In Britain, the discoveries of Galileo were popularised by, among others, John Wilkins a young puritan clergyman who became chaplain to Lord Saye and Sele, a leader of the Parliamentary party in the English Civil War. In 1638 Wilkins published *The discoverie of a World in the Moone*, and in 1640 *A Discourse concerning a New Planet* – the new planet being the Earth. When Oxford was purged of 'malignants' under the Commonwealth, Wilkins became Warden of Wadham College and gathered around himself there a group of important men of science including Robert Boyle, Robert Hooke, and Christopher Wren. He married Oliver Cromwell's sister, and was appointed Master of Trinity College, Cambridge shortly before the Restoration of Charles II in 1660. The Royalist Master who had been ejected was now restored, and Wilkins lost his job; but he soon made his peace with the new regime, and was made Bishop of Chester. He was probably the most important man behind the creation of the Royal Society, and he became its first secretary.

Wilkins, like many of his contemporaries, was disturbed at the break-up of the learned world that was resulting from the rise of nation-states and the outbreaks of wars between Catholics and Protestants. With the rise of nations went the rise of vernacular languages. In the early years of the seventeenth century Latin was the obvious language for any scholarly publication, but by the time of the Restoration science was appearing in French, English, and other modern languages. This made them more accessible (as Galileo's book showed) in one country, but much less so abroad, and the translation of science began to be an important activity –

ideas in foreign languages just do not catch on. Newton's *Principia* of 1687 was in Latin, but that was because it was a work of such abstruse mathematical reasoning that it could only be read by scholars (to whom Latin presented no problems) and needed an international market if it was to make any sense as a publishing proposition; the writings of Boyle and Hooke, much more experimental and readable, were in English.

Wilkins, who like many of his contemporaries had been interested in codes and ciphers during the Civil War and had written a book on the subject, thought that the way round this problem was to devise an artificial international language. There were different words for the various numbers in different languages, but the 'Arabic' numerals were international symbols standing for the number rather than for any of the words. Wilkins and his contemporaries had also read accounts of the Jesuit mission to China, and thus knew that although people in different parts of China could not understand each other's spoken language, the written language was common to all literates. It was generally accepted that the written characters were ideograms; that is, that they stood for things or ideas rather than for words. Those who had lived through religious wars and political upheavals were well aware of the dangers of rhetoric and of word-splitting; and indeed the Royal Society in the early days required of its members a plain style more like that of artisans than of scholars.

Wilkins hoped to devise a language which would be like the Chinese, in that symbols would stand for things rather than for words, and anybody could read it – an Englishman and a Frenchman reading it aloud would sound very different but mean the same. Not only would such a language fulfil the function which Latin was ceasing to perform, but for science it would be better than any natural language because merely verbal disputes would be impossible. As the authors of the *Vocabulary of Colloquial Navaho* (1951) put it: 'Some English speaking orators make such an effort at eloquence that, in their preoccupation, they neglect to convey an idea. Such utterances cannot be interpreted into Navaho.' The same would apply to Wilkins' language.

The Chinese symbols had perhaps begun as pictograms but had come to stand for things by convention. Wilkins wanted to base his language on a natural classification, so that there was a real connection between the symbols and the things and his script could be described (as in his book of 1668) as *A Real Character*. To accomplish this he needed to tabulate everything. The symbols would then indicate where in the table the thing was to be found, rather like a map reference. Wilkins adopted a system of classifying in threes, and he consulted with various experts about the details of the system. Among them was John Ray, who did the botanical part of the book, and Francis Willughby, who did the zoology. Ray had been at Trinity College in Wilkins' time, and had resigned in 1662 rather than take new oaths after the Restoration. He was the leading botanist of the day and had recently prepared a *Flora of Cambridge* (1660) which was the first flora of an English county. Willughby was his pupil and then his patron. After Willughby's death in 1672, Ray edited and brought out his work on fish and on birds, while he himself produced works on the classification of plants and animals. The attempt to fit plants into the procrustean system of Wilkins had convinced him that nature did not work in threes and that a return to, and then a going beyond, Aristotle was necessary if one was to get a natural system. Wilkins' artificial language went the way of most such projects, into oblivion; but Ray set taxonomists on a new road, as Galileo had done for astronomers.

The system, if one can call it such, which was in use for the classification of organisms was man-centred, and practical in that it was related to the cure of diseases or to edification; though it had some connections with a general world-view in rather the same way as the old system of astronomy had. Works of natural history were usually bestiaries or herbals. The latter descended from the great work of Dioscorides in Ancient Greece, and were designed as adjuncts of medicine and cookery. Their aim was to help the reader find plants that would be useful in the cure of diseases, or that would make a tasty dish. The grouping of things that are good to eat or that cure the same kind of illness is not going to lead to a natural system, and while in herbals one does sometimes find certain plants put near each other that are now classified in the same group, such botanical insight was not required of the herbalist.

One of the great problems was to recognise the plants described by an earlier author. A major difficulty was that the plants described by Dioscorides were from the Mediterranean region, which has a very different flora from that of Britain, Holland and Germany – it was not until about the sixteenth century that this fact became generally known, while at the same time it was becoming apparent to the Spaniards in the New World that the flora of America was very different from that of Europe. To recognise differences in more than an impressionistic way, one must have clear descriptions and generally-accepted names. To assist descriptions one needs pictures; and the pictures will be more useful as the artist comes to know which parts of the plant are the most significant for recognising it and classifying it (which are rather different requirements). Illustrations form a kind of visual language in all of natural history.

There is a Byzantine manuscript of Dioscorides, which has many recogniseable illustrations of plants; but when manuscripts were copied, perhaps by those who did not really understand them, the text might be pretty accurately reproduced but the pictures were all too likely to become inaccurate (while perhaps remaining attractive as decoration). The introduction of printing in the fifteenth century meant that good texts (if properly edited) could get into multiple circulation, and could be illustrated with woodcuts so that every copy has just the same text and illustrations. The sixteenth-century Herbals of Brunfels, Fuchs and Gerard were handsomely illustrated, described new species, and indicated what diseases the various plants might cure. The reader is amazed that any diseases survived; some plants are described as so potent that, like penicillin in the early days, they should have cured almost anybody of any illness. It was not a major part of the task of the herbalist to group plants naturally, or to take much note of their anatomy; botany remained down to the later years of the nineteenth century a science subsidiary to medicine, taught in medical schools by doctors for whom utilitarian objectives must usually have been the major consideration.

Classifications of plants in herbals were thus subjective in that they were man-centred, and depended upon the 'virtues' of the plants rather than their botanical characters. In zoology there was less emphasis on medical usefulness, though some animal preparations such as unicorn's horn were widely prescribed. The major zoological works were bestiaries, in which the point was not to give information about the nature and behaviour of various species but to illuminate human actions and destiny. The lion is the king of beasts because of his noble character, which contrasts with that of unchivalrous creatures like the crocodile. The fox is cunning; the coot looks after orphaned or rejected babies of other species; the light-shunning owls symbolise the Jews who rejected Jesus (or, alternatively, they symbolise wisdom!); the crab tricks the oyster into opening up its shell and forces in a stone so that he can get his claw in and eat up the oyster's flesh. The classification here is really into good and bad, to be copied or shunned; and while bestiaries like the famous *Physiologus* of the twelfth century do contain much truth about animal behaviour they also contain much error and there was no way of telling which was which.

Animals seemed to have been created to remind man, like stained glass windows, of the teachings of religion – and if an example happened not to have been created, then it could be invented. Thus the phoenix was a symbol of the Resurrection; but so was the butterfly, which spends its life as a caterpillar crawling around, dies into a crysalis-coffin, and rises again a splendid creature flying up into the heavens. Man's destiny was similar, and the creature's life was an allegory; one, moreover, that was popular well into the nineteenth century. Other creatures symbolised the devil: the wolf, always skulking about sheepfolds; the wild ass, always braying for its dinner; and the partridge, because of its powerful and perverted sexual appetites, and because one female may steal the eggs of another. These symbolic or allegorical classifications are clearly not natural in the sense in which we have been using the term, but they do indicate a determination to see everything forming part of an ordered pattern, which is what lies behind any natural system. Bestiaries have been well described as collections of animals with human faces; and the illustrations of most zoological works through the seventeenth century bear this out. Man's relationship with the animal kingdom was close, but it was a symbolic and not a genetic relationship. Animals were made for the use of man, sometimes directly and sometimes indirectly, to recall him through fables or allegories to the worship of God, the creator and preserver of the whole order.

In bestiaries there was sometimes another principle of order, which survived through the eighteenth century and was based upon the four elements, earth, air, fire, and water. Everything was composed of these, in the animal, vegetable and mineral kingdoms, and upon their proper proportions a man's temperament depended. More significant, from the point of view of the classifier, was the slightly different sense of element, in which the land, the air, fire and fresh or salt water formed the different habitats of the various animals. The denizens of the various elements formed parallel series; thus over the land there were the swallows and magpies that we are familiar with, and over the sea there were sea-swallows and sea-pies – terns and oyster-catchers to us. On the land there were horses, and in the water there were river-horses (hippopotami) and sea horses; there were even creatures like monk-fish and mermaids, and none of these resemblances were merely accidental. The world all hung together, the great world or macrocosm was represented in the little world, man the micro-cosm, and everything had its putpose in the divine plan.

This taxonomic scheme even had its heuristic aspect, for it could function as a working hypothesis leading to discoveries just as Darwin's theory did and does. For any creature known, one should seek for parallel creatures inhabiting different elements. The element of fire was the odd one out, but there were creatures like the fireflies that were associated with it, and the salamanders which were supposed actually to dwell in the fire – an idea that must have required some suspension of disbelief at any time. The sea-cow, discovered and exterminated in the eighteenth century off Siberia, was perhaps the last creature to be named in accord-ance with this belief in the parallel creatures of each element. There is at least a formal resemblance between this view, and the more modern one that if there are ecological niches in New Zealand similar to those in Europe, then there must be some crea-tures that fill them.

By the eighteenth century, the four element theory survived in chemistry where it was reinforced by the early work on gases of Stephen Hales in the 1720s and 30s, because these seemed to be better or worse forms of the element air. The Swede, C.W.Scheele, was led to his isolation of what we call oxygen in his investigation of the element fire in 1777; and it was not until Lavoisier in the next decade reinterpreted combustion as combination with atmos-pheric oxygen that air and fire disappeared from the list. Lavoisier also interpreted water as a compound of oxygen and hydrogen, and therefore not an element; and argued that there was a class of different bodies, probably all oxides, that had been given the name 'earth'. His definition of 'element' was intended to be an objective one; it was that substances should be supposed elemen-tary if they had never been decomposed, and was therefore pro-visional.

In natural history, the four elements survived to much the same time, but as a method of organising books rather than ordering nature. County natural histories, like those of Robert Plot who described Oxfordshire in 1677 and Staffordshire in 1686, and of William Borlase who described Cornwall in 1758, were organised within a four-element framework; and vestiges of it are discernible also in Jefferson's *Notes on Virginia* of 1787 originally drawn up in response to a French questionnaire. As an interpretation of animal habitats or of human behaviour, the four elements had become obsolete by the middle of the seventeenth century, with the publication in 1646 of the *Pseudodoxia Epidemica* of Sir Thomas Browne, an eminent physician and the author of *Religio Medici*. Despite their Latin titles, his books were in English (of a splendidly polished and latinate kind), and the *Pseudodoxia* is often given its English title of *Vulgar Errors*. In this book, he exposed the credulity of his predecessors who had believed all sorts of marvellous things about animals, and attacked the whole allegorical interpretation of zoology. While he did not himself publish any kind of classification, he cleared the ground for those who sought an objective and natural system. Descartes' view of animals as machines came to replace older ideas of the balance of elements; and Galileo's work got rid of the sphere of fire in particular, and of the four elements generally, in physics because he showed that the earth and the heavens were not composed of different substances, the four elements and the quintessence.

There was in the seventeenth century a simpler method of organsing books on natural history, and that was to put the animals in an alphabetical order. This must be artificial, since the order would be different if one were using different languages; and depended, for instance on whether one described man's long-eared servant as a donkey or as an ass. *The Historie of Fore-footed Beastes and of Serpents* by Edward Topsell which were published in 1607–8, and again with another volume on insects by Thomas Mouffet (possibly the father of Little Miss Muffet) fifty years later, were in alphabetical order, and therefore had no systematic classification beyond separation into what are roughly mammals, reptiles, and various kinds of invertebrates in the three volumes of the 1658 edition. Topsell's interests were however different from those of the earlier bestiarists, for he was less concerned with direct usefulness and with allegories, and more with the animals themselves. In this he was following his sources, particularly Konrad Gesner (or Gessner) whose *Historia Animalium* appeared in five volumes between 1551 and 1587, and who was one of the great Renaissance polymaths. His book shows as much evidence of erudition as of observation, with references to animals in the literature of Antiquity being extremely important. Indeed, to our classically-educated ancestors natural history was likely to come up in connection with Virgil or Pliny, as geography was in connection with the campaigns of Alexander the Great. The literary and rhetorical emphasis in Renaissance education, as opposed to the logical bias of the older Aristotelian scheme, did not lead immediately to the pursuit of taxonomy.

Many of the creatures described and depicted in Gesner and in Topsell are not to be found in later works; but if there was a reference in literature or in the Bible to the hydra killed by Heracles or to the cockatrice, then a description and a picture had to go in, possibly accompanied by some expression of uncertainty about the real existence of the creature. Topsell, who was a clergyman, recommended his book for Sunday reading, – 'heavenly meditations upon earthly creatures' urging that the knowledge of beasts was divine, and that of the four-footed kind especially so because they were created according to *Genesis* immediately before the race of mankind. In general, he recommends zoology because it tells us of the love of dogs, the meekness of elephants, the neatness of cats, the chastity of the turtle, and the utility of sheep. Natural history is to be preferred before human history because it is a chronicle 'made by God himself, every living Beast being a word, every Kind being a sentence, and all of them together a large history, containing admirable knowledge and learning, which was, which is, which shall continue, (if not for ever) yet to the Worlds end.' The idea that the creation resembles words and sentences forming books (which was a commonplace of atomism) is evidence that Topsell saw an ordered whole, organised by rules like those of language; an idea that was to be of some importance in the nineteenth century, but in a different way because then it was the changes of languages and animal species that were of interest while Topsell takes for granted a static world in which species endure.

Topsell makes some groupings into 'sentences', putting the roe and fallow deer together with other creatures called deer-goats; but the elk appears quite separately, and the ass is not described with the horse. He did his best to be critical, but he knew that in natural history the strangest-sounding story may turn out to be true and he usually printed it if it came in two authorities (even if one had perhaps copied it from the other). His book had an index, but it was of diseases that could be cured by medicines of animal origin and is therefore of greater value to medical and social historians than to systematists.

When the second edition of Topsell's book came out in 1658, Ray was already teaching at Trinity College, Cambridge. Like any other important scientific work, Ray's marked only a partial break with the past and he was interested in plant remedies as well as in plant taxonomy. The idea that the Moderns could see further than the Ancients only because they were sitting on the shoulders of the giants of Antiquity was a commonplace, voiced, for example, by Ray's contemporary, Newton. Ray developed and emphasised traditions going back to the time of Aristotle, and had the guidance too of herbalist and bestiarist predecessors and contemporaries. Atomism, the Copernican theory of the planets, and the revival of the natural system were among the great intellectual triumphs of the seventeenth century, and they were all rooted in the science of Antiquity and consciously developed from it. The originality lay in the choice made of the parts of ancient science, and in the way they were developed.

What was new for Ray and his contemporaries was that they had scientific societies. In Italy, disciples of Galileo had been united in the *Accademia del Cimento*; in 1660 the first steps in the founding of the Royal Society had been taken, and shortly afterwards the *Académie des Sciences* at Paris was begun, as a Department of State with salaried Academicians rather than as a private club with royal patronage as in the case of the Royal Society. At meetings of these bodies short papers could be presented and discussed; and in 1665 in Paris and in London, journals associated with the societies (but not at first formally published by them) began to appear. The scientific paper, now probably the most important vehicle for science, was born; and its circulation around the scientific community was ensured through the societies. To some extent, scientific journals superceded the corespondence by which 'invisible colleges' of men of science had kept up with each other's discoveries; but even though the paper may represent a short or preliminary investigation, it entails some formality, and correspondence went on being of great importance. Magnates of nineteenth-century science like Alexander von Humboldt or Joseph Banks wrote thousands of letters concerned with science.

Perhaps the most revolutionary works, like the *Origin of Species*, are not suited to appearance in a journal; when Darwin's brief account of his theory did so appear in 1858 nobody took any notice of it and it was not until the book came out in 1859 that the storm broke. Papers in journals are excellent for the publication of results or discoveries where the fundamental tenets of the science are not in question; what journals publish is within a tradition, or is what Kuhn calls 'normal science'. More room is often needed for radical demolition and rebuilding, or for cumulative argument, like that in a law court, which may well therefore require a book. In natural history, papers and books – presenting some new scheme, or describing a species or a whole natural group in some detail – have both played an important part in the development of classification.

Ray became an expert anatomist, and published in the Royal Society's *Philosophical Transactions* a paper describing the dissection of a porpoise that he had done when visiting Wilkins at his Bishop's palace in Chester. In the *Ornithology* and the *de Historia Piscium* of Willughby in which Ray collaborated and which he edited (he also translated the bird book), we see evidence of this ability to dissect. Aristotle is supposed to have dissected about a hundred species of fish in the course of his zoological work, and the knowledge of their anatomy enabled their family groups to be worked out much more efficiently than from external resemblances only. Bats look rather like birds, and porpoises like fish, but their inner structure is quite different. In using anatomy in classification, Ray was in the tradition going back to Aristotle.

The books on fish and on birds were large and handsome works, illustrated with engravings on copper; for illustrations where detail must be shown, these were much superior to the early woodcuts and for works of science largely replaced them. The expense of getting plates engraved was however high; the plates in the fish volume were paid for by a number of patrons, including Samuel Pepys the diarist and President of the Royal Society, whose names appear at the bottom of them. Although they cannot readily be printed on the same page as the text, as woodcuts can be, because they need a different printing press, copper plates, engraved or etched, remained the norm for accurate illustrations down to the middle of the nineteenth century. Ray and Willughby did not show dissections of the birds and fish on their plates as a rule; the fish got a plate to themselvs in most cases, but the birds are grouped so that each plate shows a number of related species.

We should note that terms like 'genus' (relation) and 'family' have been used by naturalists since Aristotle without necessarily any commitment to belief in a real genetic connection between 'related' species.

The *Ornithology* is arranged systematically, so that birds of similar kinds are described one after the other, as in a modern work, although the groups are not exactly the same. Exotic birds are included; one heading is 'The red-breasted Indian Blackbird, perchance the *Jacapu* of Marggrave'. They had seen this bird in Tradescant's 'Cabinet', one of the famous collections of rarities of all kinds of the seventeenth century; and the heading illustrates a problem faced by systematists well into the nineteenth century. Willughby and Ray had only the 'case' of the bird to go on, and for identifying it they had to rely on comparing this mummified object with descriptions in a literature where a precise technical vocabulary had not yet been developed. The long lists of synonyms in works of the early nineteenth century show that this was not an easy problem to solve. Many of the illustrations in the *Ornithology* look very stiff; this is because they were done from long-dead specimens of species quite unknown to the artist and without any field sketches.

Willughby and Ray put the swift among the swallows, as was generally done right through the eighteenth century, although they had done some dissections in preparing their descriptions. But of the sea-swallow which Aldrovandi, another Renaissance polymath, had put among the swallows, they wrote 'This bird, in my judgement belongs not to this Family, but ought to be ranked with the lesser *Lari* or *Sea-Gulls*'. The full description, including dissection and an account of its nesting-sites, is duly to be found among the gulls. Immediately below it comes the 'lesser Sea-Swallow', of which it is related that 'Their flesh is good to eat'; early naturalists often did a speedy dissection and description before whatever they had caught was put into the pot and a type specimen thus lost for good. Natural historians and chemists were less squeamish than we are, and the taste of things (animal, vegetable or mineral) was recorded as a matter of taxonomic importance.

In classifying the tern among the gulls rather than the swallows, the authors made it clear that this was a matter of judgement. Any classification, or any theory, involves this element, and that is why it is wrong to think of the sciences as masses of authenticated facts; the ordering of the facts is the important part (and without some ideas of the order of things one could not even collect them), and this is something related to other creative activity – like the sculptor producing an ordered shape from an enormous stone. However, scientific classification is not pure judgement, but educated judgement; the best way to learn classification in the seventeenth century would have been to work with Ray or one of his eminent contemporaries, and to take up where they left off. In learning to ride a bicycle, descriptions of what to do are not much help; one just has to leap on and try, and one finds that the falls become less frequent with time, especially if there is somebody helpful to hold on to the saddle. Classifying (and most things we do – the most important things that have to be learned cannot usually be directly taught) is a bit like this, but there are rules and Ray drew up some in his *Synopsis Methodica*, the mammals and serpents volume of which appeared in 1693 and the birds and fishes posthumously in 1713. These set out formally the system Ray used, no doubt making it tidier than it really was – Einstein said that one should look at what scientists do rather than what they say they do, and Ray's system looks more a matter of bifurcating divisions and less a matter of judgement of families when he sets it out in his *Synopsis*.

In both the *Ornithology* and the *Synopsis*, despite any slight differences in emphasis, Ray was trying to place creatures in their family groups; and he was relying on multiple criteria, though one might be dominant – in classifying mammals, for example, he relied upon their feet to form the major classes. Although he alluded to the man-like appearance of the monkeys, he made no attempt to classify man among the mammals. His contemporary Edward Tyson, the physician at Bedlam Hospital, the famous lunatic asylum, published in 1699 his *Orang-outang*, whose Malay name he translated into Latin as *Homo sylvestris*, the wild man of the woods. This was perhaps misleading, because he did not believe that the ape was a kind of man at all, but a species intermediate in the scale of nature between men and monkeys. The animal he was describing was really a young chimpanzee, and his was the first monograph on an anthropoid ape and remained a standard work for over a century. In Ray's day, even after Tyson's book had appeared, there was no need to place man among the animals; indeed, Tyson had separated him, although pointing to the many similarities.

Ray had come to zoology from botany, a science which had

moved ahead of zoology in terms of efficient classification and was to maintain its lead through the eighteenth century. In botany, Ray began with the crucial distinction between monocotyledons, which produce a single seed-leaf, and dicotolydons, which produce two seed-leaves; though whether he saw this distinction as a fundamental, natural one or merely as very convenient has been doubted. He emphasised in his *Methodus* of 1682 the importance of taking into account all the characters of a plant in classifying it; not just the seeds or the fruit, but also the flower and the leaves. The difficulty with such an approach is that it makes classification such a demanding business that it can only be done by experts who have large collections at their disposal for comparisons.

In fact, Ray kept some of the traditional categories because he was engaged in a practical as well as a theoretical endeavour, and they were convenient. From Theophrastos, a pupil of Aristotle, had come the idea of separating plants into trees, shrubs and herbs; a classification useful for the gardener, but one which cuts across natural groups – a relative of the common weed groundsel, manifestly a 'herb', grows on the top of mountains in Africa into something looking like a very curious tree. In the eye of modern botanists, therefore, many of Ray's families are not natural groups, and in this field his theory was perhaps more modern than his practice. He recognised that a natural system could not be the product of a single botanist, or even of a single epoch.

In Ray's time, recognition of the sex of plants was only slowly coming, and Tournefort, a very important younger contemporary of Ray, doubted it. This was something where science was to generalise something that must have been long known to fruit-growers, especially those concerned with dates or figs. The sexual parts of plants were therefore not made much of in classification in the seventeenth century, though they were to come into their own in the eighteenth century as the basis of Linneaus' system. The anatomy of plants was also little regarded in Ray's time, although a microscope had been demonstrated by Galileo at the period when he was making his discoveries with the telescope. But by the 1660s, the microscope was being applied to plants, and the two great pioneers of plant anatomy were Malpighi and Grew, working especially during the 1670s. The magnificent illustrations of Grews *Anatomy of Plants*, cut away to give the impression of three dimensions, remained standard well into the nineteenth century, simply being copied for example in Humphry Davy's *Agricultural Chemistry* of 1813; but it was a long time before anatomical detail was taken into account as a matter of course in botanical classification.

By the time Ray died in 1705, the essentially artificial or subjective systems that had been in use a hundred years before had been undermined. Groups or families of creatures were recognised, and species were being placed in them; but the task was a difficult one and there was little agreement on the exact rules to apply – mainly because in a matter of judgement like forming classes there were no exact rules in *the nature of the case*. Since plants and animals unknown to science were being found in large numbers in all parts of the world, the job of naming and classifying them was a daunting one. Ray and Willughby's problem, of deciding whether an 'Indian Blackbird' was the same as the 'Jacapu' described from Brazil, was similar to that faced over and over in the eighteenth century. The solution was to adopt a system based upon one fundamental character, which would therefore be artificial but would be easy to apply. If the character chosen was really fundamental, then the classes so determined would not differ too much from natural groups. The step to a system of this kind was taken by Linnaeus; and in taking it, he brought it about that natural history (seen as essentially a descriptive and classificatory science) became the leading science of the eighteenth century, as astronomy had been for much of the seventeenth. Unlike Ray, Linnaeus did not hesitate to classify man too, putting him with the monkeys in the Anthropomorpha division of the Quadrupedia in the first edition, of 1735, of his *Systema Naturae*, and instead of an enumeration of toes to describe the group, putting 'Know theyself'. It is to Linnaeus' period that we must now turn.

3

The Artificial System

RAY'S FRENCH ADMIRER Tournefort carried on the work of botanical classification, and worked out a number of families which are still regarded as natural groups; in the first third of the eighteenth century, his was the leading system, and it was intended to be a natural one. Because such a method of classifying demanded that a range of characters be taken into account, it was laborious, and it was very difficult to do in the absence of collections of dried plants, of a botanical garden – a necessary part of the equipment of any respectable university by the middle of the seventeenth century – and of expert colleagues. All these requirements could be met in the great centres of learning in Europe, where there were the 'cabinets' of wealthy and eminent persons as well as universities; where magnates competed in the naturalisation in their gardens or their heated greenhouses ('stoves') of exotic plants; and where there were scientific societies and academies meeting regularly for the exchange of information and advice, and publishing their transactions. In medical schools, and especially in the early eighteenth century at Leyden where the great Boerhaave was a professor and was in charge of the botanical garden, instruction in botany and zoology and chemistry formed an important part of the curriculum; and when the Edinburgh medical school was begun on the Leyden model, classes in botany were given there. Taken through the natural system with a tutor, one might hope to acquire with practice the intuitive grasp of natural groups, so that one could place a plant in its family taking a whole range of characters into account.

By the early eighteenth century, most of the more obvious plants of northern Europe had been described, and there was not much hope of finding a 'non-descript' flowering plant within walking distance of Leyden, Edinburgh, or Paris. It was in more remote regions that discoveries were to be made; and for much of the eighteenth century the emphasis in botany was on discovery of new forms, rather than on the close study of geographical distribution and variation which became characteristic of the nineteenth century. North America was an area which was botanically (and zoologically) little known, and which was by 1700 settled along its eastern margin by colonists who now had sufficient wealth and leisure to begin to take a strong interest in the sciences. The sciences have always proved much easier to export than the fine arts; those brought up without the chance to see the paintings or drawings or sculptures of the masters, and without access to good libraries, have little hope of achieving easy familiarity with this kind of arts-based culture, whereas for American colonists or Manchester manufacturers or Japanese samurai the sciences (and especially natural history) were an open door.

To work the natural system, however, one needed something like the connoisseurship which enabled a man who had been on the Grand Tour and had carefully looked at many paintings to be sure that a given painting was by Titian, for example. If such an expert is challenged – and experts then and now do differ, and make mistakes in these kind of judgements – he could not appeal to definite rules, but would list a number of features which a painting by that artist would be expected to show. An outstanding virtuoso or collector or advisor to a nobleman might add more diagnostic characters, and thus separate paintings by the great artist from those by his most successful pupil; but his techniques could only be learned by an apt pupil being guided through a large number of examples by an expert tutor. This was something available in natural history to the keen student at Leyden or elsewhere, but not easily obtainable in the colonies; and, in any kind of connoisseurship, the dictum applies that the self-educated man will have had a very ignorant instructor. What was needed was a system that was like the plants, cut and dried.

The problem with the natural system was even worse in America or in the Far East where doctors and factors with the East India Companies were beginning to take an interest in the fauna and flora of the places to which they were posted, especially as this became an increasingly gentlemanly interest. Just as wide knowledge of European painting would not immediately open the door to understanding and appreciation of Japanese art, so the natural groups of Europe were not necessarily represented in other regions, and the intuition of the expert might be baffled for some time. Even he would then welcome some definite rules, so that one could rapidly give names to the new plants or animals and as it were index them, and later get around to putting them into their natural relations. The rules were supplied by Linnaeus, who thereby made natural history a manageable science for the rest of the century after his *Systema Naturae* first appeared in 1735.

Alexander Garden was a physician trained at Edinburgh who went to Charlestown, South Carolina, in 1752; and three years later he wrote to Linnaeus, seeking to begin a correspondence with the great man who was becoming the dictator of natural history. His letter showed the untruth in science, as elsewhere, of the idea that flattery will get you nowhere; but the point that he made about Linnaeus' system as compared to any that had gone before would have been generally agreed, at least outside France, in the second half of the eighteenth century. Garden wrote that he had thrown away two years,

> In following Tournefort's system, in which I was first instructed by Dr. Charles Alston in the Edinburgh garden. Having been invited to Carolina (where from that time I have, with tolerable success, practised medicine), being furnished with the *Institutiones Rei Herbariæ* and the writings of Ray, I made daily excursions into the country. But the immense labour of reducing the plants I collected into proper orders, to say nothing of the uncertainty attending the investigation of genera and species, and, still more, of determining varieties, according to the Tournefortian system, was all so very tiresome, that at length my patience was exhausted; and had I not, by good fortune, met with your most excellent works above-mentioned, I might have been stopped in my progress, and have altogether given up this most pleasing of pursuits.

Linnaeus duly entered into correspondence with Garden (naming the *Gardenia* after him) who sent him specimens of animals and plants from North America which, being described in Linnaeus' works, have become the 'types' of their species and genera.

In another letter, Garden wrote in 1760 to Linnaeus that 'although the wonderful works of God were constantly before my eyes, while I was destitute of a true guide I viewed them all to no purpose'; phrases of this kind, indicating the hope of going from Nature to Nature's God were commonplaces of natural history from the days of Ray (whose most famous book was *The Wisdom of*

God manifested in the Works of the Creation, 1691) down to the middle of the nineteenth century; we shall have to look closely at some specific examples of natural theology later, but it was ubiquitous. Where there was no obvious use for a plant, an animal, or a mineral, it at least disclosed something about the Creator; collecting in natural history was not a frivolous pastime – the 'stamp-collecting' of Rutherford's epigram – but a way of gaining insight into how God worked.

Perhaps less characteristically, Garden wrote to another of his friends in science that treatises 'done in the systematical order' would be very acceptable, adding 'I think, if a system be any thing clear and ingenious, it is by far a greater help to the memory and judgement, than all other helps of descriptions, prints, cuts, drawings, &c. 'It seemed to him that any ingenious system would be based on some essential character, while an illustration, giving merely an external appearance, could be thoroughly misleading. He was in this spirit censorious about the illustrations in Mark Catesby's *Natural History of Carolina*, 1731–43, the pioneering work on the area (much used by Linnaeus), generally admired at the time, and since, for its large and spirited plates etched by Catesby himself. Garden found inaccuracies in details where most of those who saw the plates were looking for the general effect; but it was Garden's concern with details, and especially his demand for system, that perhaps mark him out from the virtuosi who were the patrons of natural history on the one hand, and from the collectors who sold specimens, cuttings and seeds to wealthy collectors or to nurserymen on the other. Garden was only a part-time man of science, but he was not an amateur.

In Linnaeus' writings, Garden found a system that was both workable and straightforward (could be picked up from books), and which seemed to disclose the order which God had imposed upon the world. Linnaeus was born in Småland in Sweden in 1707, the son of a clergyman; his mother had wanted him to enter the Church too, but to her disappointment he did not do well enough to get on to the theology course and took medicine instead, first at Lund and then, after a year, at Uppsala. Both universities had very small medical faculties, and the few professors were not very active. Linnaeus worked under his own steam in the botanical garden and the libraries of Uppsala, and attracted the attention of Olof Rudbeck jr., the professor of botany, who appointed him as a temporary 'docent' or assistant lecturer with the duty of demonstrating the plants in the garden to the medical students.

In 1731 the temporary appointment came to an end, and Rudbeck persuaded Linnaeus to plan a journey into Lapland which the Royal Scientific Society at Uppsala agreed to support – thereby doing in its small way what the bigger scientific societies in London, Paris and elsewhere were to do frequently during the eighteenth century. The journey was a milestone in Linnaeus' career, for he accumulated much information and began to think about how to organise the data of natural history. In 1735 he became engaged to the daughter of a physician, and set off for Holland to complete his medical education, as many Swedes (and other foreigners) did. He took with him manuscripts which were the foundations of his major works, but he came as an unknown foreign student. He visited Amsterdam and Leyden, but took his degree at the smaller and cheaper, but respectable, university of Harderwijk; the whole process of enrolment, passing the preliminary examination, getting his thesis (which he had already written) printed, attending some lectures, and passing his oral examination, took a week.

Now a qualified doctor, he went to Leyden and met the eminent scientists Boerhaave and Gronovius, who were very impressed with his abilities and his manuscripts, and who became his patrons. Gronovius saw to the publication of the *Systema Naturae* in December 1735; and Boerhaave arranged for Linnaeus to be made physician to the banker George Clifford who at Hartecamp, his country house near Haarlem, had made a magnificent garden with all sorts of exotic plants. Linnaeus produced a splendid volume cataloguing the garden, and, while in Holland, also published his *Genera Plantarum, Flora Lapponica*, and *Critica Botanica*. He never learned any Dutch, Swedish and Latin being his two languages, but his friends in Holland tried to persuade him to stay there, hoping to arrange a chair of botany for him; but he was determined to return home, and in 1738 he went back to Sweden. During the three years he had also visited England and France, meeting eminent natural historians in both countries. In 1739 he married, having established a medical practice in Stockholm, and in the same year was a founder-member of the Royal Swedish Academy of Sciences, and its first President. In 1741 he was appointed to a chair at Uppsala, and apart from some short journeys remained there until he died in 1778.

At Uppsala he trained numerous pupils, twenty-three of whom themselves became professors and helped to spread his system, while others travelled to remote places enriching the Linnean collections and describing their finds in Linnaeus' nomenclature. In his youth, students wanting to get a good education in natural history had gone to Holland; he reversed this process, and attracted pupils from abroad. He seems to have been an excellent teacher; the eighteenth century was the age of the encyclopedic ideal, and Linnaeus fitted in excellently into such a tradition. He was prepared to take great pains over the organisation of material, and its presentation in a convenient and accessible form – he may be seen as a pioneer of information retrieval. Those who look for great originality of mind in a great scientist find themselves disappointed with Linnaeus; his work was not profoundly original, but a synthesis of ideas and practices that had been used by earlier naturalists – and an orderly synthesis was just what was needed at that time. There would have been no scope for a Newton of natural history.

Linnaeus had begun with the genera, having published the *Genera Plantarum* in 1737 while he was in Holland. His genera were definite taxonomic units, groups of species, rather than the vaguer family groups to which the term had been applied in Antiquity. Right through his classification there is a tension between a deductive kind of system based on some 'essential' character of each group, and an inductive one based on multiple characters: he recognised that his system was not natural, but he did believe that some of the groups at least were natural. For plants, he developed the sexual system, dividing plants into twenty-three 'classes' based on the number and relations of the stamens of the flower, such as the *Monandria* and the *Pentandria*, with one and five stamens respectively. The classes were then divided into 'orders' depending upon the number of styles or stigmas, so that *Digynia* for example has two. The genus *Syringa* then belongs, for example, to the *Diandria Monogynia*. The determination of the class and order to which a plant belonged ceased for followers of Linnaeus to be a matter of long training and of judgement; it simply depended on counting, and should be straightforward without borderline cases.

In such a system it would have been possible that some combinations might never have been found, so that there might be some vacant places in the table. This did not happen with

Linnaeus' botanical system, but at the turn of the century Lacépède in France did have some vacant spaces in his classification of fish, and in chemical classification the gaps in tables proved to be highly significant. It is only when numbers enter into classification that this possibility arises in a serious way, though anybody interested in family relationships could perhaps design a hypothetical cousin in whom some possible grouping of family traits was found that had not so far been realised. In Linnaeus' system the trouble with this clear and numerical division of the major classes was that it brought together plants that to the practiced eye of the expert botanist had little in common, and separated others that were very similar in most respects. It was a convenient system, which could be quickly learned by a ship's surgeon before a voyage to some exotic region, or by anybody with an interest in wild flowers or in gardening – which then, as now, was a large group of people.

The system was sexual, and would not have been thought of had the sexual nature of plant reproduction remained unknown; but it did not depend on any close study of mechanisms of fertilisation. The numbers of the sexual parts are no longer believed to be especially significant, and Linnaeus' sexual system is not fundamental as Aristotle's emphasis (in his zoology) on methods of reproduction was. But the idea, explicit in Linnaeus' names for the classes and orders, was that the plants not merely married but went in for all sorts of polygamy and polyandry, like 'twelve to nineteen husbands in the same marriage', 'husbands live with wives and concubines', and 'many marriages with promiscuous intercourse'. The language of flowers had long been used between lovers, but suddenly it became rather more explicit. There were some who objected to this intrusion of lewd imagery into science, but the eighteenth century was not a period of great prudery and the idea that the plants had sex lives made the Linnean system more popular and did not prevent its spread among the ladies. In the nineteenth century, John Ruskin, who used to make art students draw plants as an exercise, made an incursion into botany; he proposed a new nomenclature that would get away from the associations of the Linnean one, but was thoroughly ridiculed for his pains – a century after Linnaeus, an art critic, however eminent, could not expect to shake the foundations of botany. Ruskin's drawings, though attractive, did not show the detail necessary for classifying; and his objections to sexual lanaguage seem silly or unhealthy in one who had sensible things to say both about art and about the social system.

When it came to the genera, Linnaeus' great object was to stabilise them as far as possible. Different authors had produced different groupings of species, giving different weight to various characters. Genera involve the recognition of similarities, and species within them the recognition of differences. At all times there have been 'lumpers' who pay little attention to small differences and try to keep the number of species and genera down, and 'splitters' who feel that no differences should be neglected and who therefore see the need for many more groups. This was not the problem for Linnaeus so much as the impermanencc of genera, so that the place of a given plant in a system was not fixed and it was difficult to know whether two authors were writing about the same plant or not. Sorting out synonyms was a major task for the natural historian in the eighteenth and the early nineteenth centuries, and indeed since then.

It is for his nomenclature that Linnaeus' works have survived, and are many of them in print in facsimile editions. In the seventeenth century there were not even vernacular names for all the plants and animals of Europe; Mme Merian, who is justly celebrated for her pictures of the butterflies of Europe and of Surinam, could find no names for some of the smaller European species. Although, as Alice told the gnat, the insects did not answer to their names, it was clearly necessity for the zoologist to name them so that others would know which was which. In the wood where things had no names, they had no character either. There were general terms, like 'sparrow' in the King James Bible which is supposed to mean any small bird, while 'lily' applies to any wild flowers growing in the fields. In vernacular languages, plants and animals were and are often given a binominal name: the red oak, the English oak, and the holm oak, the goat willow and the crack willow, the blue tit, the great tit and the coal tit being examples. In these cases the terms 'oak', 'willow', and 'tit' correspond to the genera, and the adjective gives a more or less informative specific designation. Where there is no very similar plant or animal, then the well-known species will be given a single name, like the wren – until similar birds are discovered, like the blue wren of Australia. The wren here would have corresponded to a monotypic genus – one which contains but one species – and even when others are discovered the original one continues without a specific qualifying term. The elm and the wych elm, and the ash and the mountain ash, would be other examples; though the latter represents a mistake, in that the two trees have certain features in common but to the botanist belong in quite distinct groups.

Like the Kalam of New Guinea, then, we are accustomed to using a system of naming not altogether different from that used in science. In the seventeenth and eighteenth centuries, the scientific name of a plant or animal came to be a Latin sentence; the world of natural historians was perforce an international one, and many of its leaders were physicians to whom the learned tongue was perfectly familiar from their universities where lectures, disputations and theses were in Latin. The description that thus corresponded to a name could become very wordy; and by the eighteenth century there were an increasing number of collectors and gardeners who lacked a classical education, but were nevertheless competent botanists. Linnaeus drew up rules for terse and orderly descriptions, imposing limits on the number of words permissible, and thus both removing confusion and saving space.

It is not easy to fit Ciceronian Latin to the exigencies of something like a telegram, where the maximum information must be got across in the minimum number of words; and Linnaeus in effect coined a new language, botanical Latin, in which verbs are almost dispensed with but the range of adjectives (carefully defined) is wide. Whereas Wilkins had tried to devise a new language from scratch, an artificial language for science, Linnaeus pursued the more promising course of adapting an existing language for his special purposes. Botanical Latin has continued in use for plant descriptions ever since Linnaeus' day, which as an eminent botanist has pointed out is a longer period than the 'golden' and 'silver' ages of classical Latin.

Although Linnaeus thus simplified Latin into a technical language for naturalists, he had the pride of the self-made scholar in his erudition and his Latin, and looked down on the vernacular languages even of Europe as vulgar and barbarous tongues. He demanded that names must come from Latin or Greek, or at least must look as though they did, and his successors found themselves having to consult classical scholars. Generic names Linnaeus coined either directly from the ancient languages, or from the names of eminent naturalists – he was not willing to let otherwise famous persons stand for a genus, but did occasionally break his own rule in favour of patrons of the science. Many earlier generic names were ruled out by the application of these rules, often to the annoyance of contemporaries: and various plants which belonged to places in America or in Asia that the Romans never knew received classical flower-names. Classical Latin or Greek may therefore be misleading when one is identifying plants. Also ruled out were names taken from local languages. Linnaeus' contemporary, the Frenchman, Adanson, went to Senegal and described the plants and shell-fish of West Africa, publishing their names in the Wolof language, and was indignant when Linnaeus renamed them – he was not even mollified when Linnaeus christened the baobab tree *Adansonia* in his honour.

Linnaeus insisted upon single-word names for the genera, but in his early writings he did not use a binomial system, and even at one time resolved to prohibit specific names. The species within the genera were distinguished by terse epithets, describing the essential features of the plant, or at least giving a 'diagnosis' enabling it to be identified, in as few words as possible. He first used the binomial system right through a book in his *Species Plantarum* of 1753, which has since been made the starting point for valid botanical nomenclature; even binomial names used before then are not valid. The system had first been tried in the index to one of his minor works, and the convenience of indicating a plant by two words meant that the convention was promoted from the index to the text. Usually a key word from the description was used as the specific or 'trivial' name; thus the red oak is *Quercus rubra*, the English oak *Q. robur*, and the holm oak *Q. ilex*; but unlike previous binomial names, Linnaeus' were not really intended to be extremely brief definitions but to be keys to a system. Within twenty-five years their convenience had ensured their success; for the oaks, for example, they are still in use and the binomial name should properly be followed by 'L.' as an indication that Linnaeus named it. Plants named by other botanists have an abbreviation of their name after the binomial; and if the classification has been revised so that the genera have been changed from Linnaeus' (or another original namer) then his initial appears in brackets: thus the grey alder is *Alnus incana* (L.), because Linnaeus had put it in the same genus as the birches, *Betula*. Other botanists, like users of natural languages, had employed binomial names from time to time, but Linnaeus saw the value of using them exclusively and consistently even in monotypic genera. Short names, published with sufficient but brief descriptions to which they were the key, and with the sexual system to get one to the right part of the vegetable kingdom, made botany an orderly subject.

Linnaeus was aware of variation in nature, and his large collections of specimens and of illustrated books and journals provided him with the 'type' specimens, against which others could be compared and sent to other botanists as duplicates, or could be judged to belong to a different species (or, later, sub-species – this last taxon was not a feature of the Linnean system). In the last resort, therefore, whoever controlled Linnaeus' collections controlled naming; the publication of his books did not supercede the use of the original material, which for botany was mostly pressed flowers annotated by Linnaeus or one of his students. The building up of big collections, at first in private hands but increasingly in public museums, was a feature of the Linnean period, and for classifying and naming it was essential.

At his death in 1778, Linnaeus' collection passed to his son (though this was contrary to Linnaeus' will, and his son only got it after it had been for a time locked up, exposed to mould and rats); and at his death in 1783 his mother offered it for sale to Sir Joseph Banks, who had previously expressed an interest in it and was now President of the Royal Society. When the letter came he happened to have invited James Smith, a wealthy young man with a keen interest in natural history, to breakfast, and since he no longer wanted the collection himself he suggested to Smith that he buy it. He did so, for a thousand guineas, and in October 1784 it arrived in London; the King of Sweden was sorry to see it go, but the story that he sent a ship in pursuit of the one carrying it to England is a fiction, although there are spirited pictures of this non-event (as of that other fictional event in the history of science, Galileo dropping things from the Tower of Pisa). After making a Grand Tour, Smith in 1787 returned to London and in 1788 he set up with others the Linnean Society of London, the first major society in Britain devoted to a single branch of science, in this case natural history. When Smith died in 1828, the Society bought the collection for three thousand guineas and it has remained there ever since. Some smaller parts of the collection had been dispersed at various times, and there are Linnean specimens in Sweden and elsewhere too.

Like anybody classifying, Linnaeus tended to take it for granted that he was ordering stable entities. If everything is in flux, and the world proved very different in the present from how it had been in the past, then science would be impossible. We now have come to accept that species do change over time, but the idea only became acceptable when there were some laws and mechanisms that made it orderly instead of a mere flux, and also when a length of time undreamed of by most people in the eighteenth century was made available by geologists. At first, Linnaeus believed, not just implicitly but explicitly, in the fixity of species. He recognised some variation within a species, but the varieties that caused more difficulties were those produced, he believed, by interbreeding of two species. True species went back unchanged to the Creation, while varieties had been generated since; often by man, improving fruits or vegetables, and seizing upon favourable crosses to breed new strains. As he got older, Linnaeus became less certain about the fixity of species, and came to think that only broader groups had been originally created, and that hybridisation had over time given us our species.

Linnaeus' idea of a species was therefore morphological; that is, it depended upon the shape and form of the organism. The sexual system for plants was not based on a study of function; and Linnaeus did not use the plant anatomy begun in the previous century by Malpighi and Grew. It was the outside of plants rather than the inside that enabled one to place them in the system; since the classifying was often based on a dried specimen or perhaps even a picture, this was a fortunate aspect of the Linnean system. His contemporary and correspondent John Ellis, a London merchant who was a patron and practitioner of natural history, used chemical tests as part of his proof that the corallines were animal rather than vegetable; but Linnaeus did not fully understand Ellis' researches, though he admired him, and did not himself make use of such tests as a guide of any importance in classifying.

Biologists do now use chemical tests, looking at the composition of egg-whites when deciding if birds belong to the same natural group, for example. Even in his own day, Linnaeus' definition of a species ran up against the idea current in zoology that a species was a collection of interfertile creatures. Hybrids were possible, for the mule was a familiar example; but mules were infertile, and so a new species of mules could not be brought about by hybridisation. Every mule had to be bred from a donkey and a horse. Creatures might look rather different, like the various races of dogs do; but it was not morphology but interfertility that was the determining test of whether they belonged to the same species. The great upholder of this view was Buffon, whose influence in France ensured that the Linnean system did not take root there as firmly as it did in Britain.

Linnaeus was primarily a botanist, and it was in botany that his system was most successful. The *Systema Naturae* included the three kingdoms of nature: the animals, the vegetables, and the minerals. The book contained the famous aphorism, '*Minerals* grow; *Plants* grow and live; *Animals* grow, live and have feeling. Thus the limits between these kingdoms are constituted.' The work of Ellis made this tidy boundary between the animals and vegetables harder to maintain. The first edition was a slim volume with tables of the three kingdoms, and aphorisms both of a general kind and also referring to each kingdom. As far as mineralogy was concerned, Linnaeus tried to find a logical system; but he was unfortunate in supposing that all rocks derived from the '*primary*' soils, sand and clay; and in referring to '*petrifactions*' as the delight and temptation of several modern authors who had made unnecessary genera, when in fact there were only seven genera and no more were possible. The ordering of minerals turned out to depend upon mathematical analyses of crystal forms, and upon chemistry, and Linnaeus' system, lacking the convenience of his botanical one, never became of much importance.

In zoology things were rather different, and valid zoological names also go back to Linnaeus, to the tenth edition of his *Systema Naturae* of 1758. By this time the tables of the first edition had disappeared, so the classification cannot be seen at a glance – but it will fit into a book of ordinary size. Linnaeus' patron Gronovius wrote in a letter of 1738: 'by his Tables we can refer any fish, plant, or mineral, to its genus, and, subsequently to its species, though none of us had seen it before. I think these Tables so eminently useful, that every body ought to have them hanging in his study, like maps.' The reason why he confined his zoological reference to fish is probably because Linnaeus' fellow-student Peter Artedi had shown the same interest in classifying; he and Linnaeus had indeed worked out their ideas together, and Artedi's great interest was in fish. When he died young, Linnaeus brought out his

Icythyologia in Holland in 1738; and the fish were thus the first group of animals to be organised in a 'Linnean' manner – though Linnaeus was explicit in acknowledging his debt to Artedi.

In the first edition of the *Systema Naturae*, the 'quadrupeds' were distinguished into orders according to their teeth, the birds according to their bills, and the insects according to their 'antennae, wings, &c'. Teeth and bills correspond to a kind of ecological division, being in correspondence with what the animal eats; and teeth remained extremely important for the classification of mammalia from Linnaeus' day on. They were especially useful for fitting fossil creatures into the scheme of things, for teeth are more likely to survive and become fossilised than other, softer, parts of the skeleton. But a system based upon teeth alone cannot be other than artificial. The zoological Table sets out the families, the orders, the genera, and the species, as for plants, with terse generic characters; the names of various genera, such as 'Bos' also being those of representative species within the genus, for species of Bos include Bos (Cows) and Bison. Zoological names still differ from botanical ones in that it is allowed to have this kind of repetition, the wren for example being *Troglodytes troglodytes*. In 1735, Linnaeus had not yet worked out his binomial system, so that although the Table seems to imply binomial names these were not really a feature of it. The Table also included a list of 'paradoxa' like the unicorn and the phoenix, which Linnaeus exploded. His invertebrates make much more curious reading than his vertebrate lists; for under the insects, in the order *Hemiptera*, we find, for example, grasshoppers, ants and scorpions, while the order *Aptera* contains fleas, crabs, and centipedes. The family of *Vermes*, or worms, included creepy-crawly creatures like snails and slugs in an order called *Reptilia* (our reptiles coming under *Amphibia*, along with frogs), the shellfish (*Testacea*), and then an order called *Zoophyta*, creatures on the boundary between plants and animals, generally soft species like sea-slugs, octopus, and starfish. To sort out the *Insecta* and the *Vermes* which were the rag-bags of the Linnean zoological system, was to be one of the great tasks facing his successors.

By 1758 when the tenth edition of the *Systema* appeared, zoology occupied the first volume, a fairly hefty octavo tome of 800 pages rather than a Table like a map – zoology had grown too complicated to be mapped in quite that way. Linnaeus knew of nearly 4,400 species of animals – a fraction of those known to us, because the invertebrates were still so little explored, but a formidable total to classify. His zoological system, compared with earlier classifications, showed the same virtues as his botany: it was concise and consistent, and the binomial names which were by 1758 a feature of the system proved very convenient.

Linnaeus had by this edition come to use the term 'Mammalia' in place of the 'Quadrupedia' of the first edition; this not only made the place of man, the biped, more obviously with the *Simia* (where he had in fact been in the first edition), but also meant that the footless whales could come out of the fish into the mammals. Reptiles now had their modern sense, approximately, and *Mollusca* had been introduced as a term for the *Vermes* and *reptilia* of 1735. The volume is set out with the general heading, such as INSECTA APTERA, with the genus being described, such as Cancer, following it in lower case type, running along the top of each page. When we first meet the genus, there is a terse description; for the mammals, this is simply a list of the teeth. Underneath, there follows a list of the species, each briefly characterised with reference to a few authorities cited in standard abbreviations, with an indication of habitat (such as 'Asia' or 'America') and perhaps another brief remark. The genera and the species are numbered; the binomial name for a species consists of the generic name, appearing as a heading or running head with the order, and the specific name which appears in the margin beside the description. For animals as for plants, there was a system which (for vertebrates, at any rate) allowed anybody to name a specimen without extensive training.

To make large collections and get them into order did require proper training and knowledge of the literature; and Linnaeus' pupils played an important part in spreading these in the second half of the eighteenth century. Men of science like Sir Hans Sloane, and patrons of science and horticulture like Peter Collinson the wealthy Quaker woollen-draper, had commissioned collectors in, or sometimes sent out a collector to, foreign lands, especially to North America. The Dutch had done the same, and the botanical garden at Leyden and Clifford's garden were rich also in oriental species and those from the Cape of Good Hope. Linnaeus had turned down the suggestion that he should go on a voyage to the Tropics when he was in Holland – he wanted to go home and get married. But as the number of keen students of his methods at Uppsala increased, so he sent out his protégès around the world.

Peter Kalm for instance went to North America, Falck into Siberia, Solander and Sparrman with Cook into the South Seas, Thunberg to South Africa and on to Japan and Löfling and Rolander to South America. Several died on their voyages, as martyrs of a kind to science: Forsskal in Arabia, Tärnström in Indonesia, and Berlin in West Africa among others.

The professionalization of science is something which has been attracting the attention of historians recently, and one of the great examples cited is Liebig's transformation of chemistry. From 1825 Liebig undertook the training, in a laboratory, of large numbers of research students for PhD degrees. He could do this because he had perfected the techniques for the analysis of organic substances, inventing, for example, the Liebig condenser so that volatile substances no longer escaped into the air. Organic analysis had previously been a very difficult business; now it became almost routine, so that Liebig's students could analyse some natural product in a relatively brief time, and earn their degree. He also controlled a journal, so that their work could be published. Linnaeus' position was not altogether dissimilar. He had perfected a way of naming and classifying species, making it almost routine, and he attracted many students who wanted to know how to do it. They did theses for the MD degree; many of them, like Liebig's students, themselves became professors and spread their master's ideas and techniques among the next generation of students. Their work would find public recognition in the successive editions of Linnaeus' works, and perhaps also in journals; the theses published at Uppsala were not written by the students, but were papers of Linnaeus' which the student would defend at a public disputation to establish his mastery of the subject.

To have studied under Linnaeus opened numerous doors; his pupils were in demand as naturalists on voyages or travels, and might hope for posts in universities or academies, or directing a botanical garden. But there was not as yet a career structure in natural history; European society worked by patronage, and a grandee like Sir Joseph Banks, President of the Royal Society from 1778 until 1820, owed his power as much to his social position as to his scientific eminence. In the eighteenth century there are some features of the world of science which seem modern, and others that seem archaic; and it is anachronistic to speak of the professional scientist in that epoch, even if it may later become useful. One major reason for this is that Linnaeus' pupils were only one group among the 'Linneans'; others were his older or younger contemporaries, who had met him in Holland or in Sweden, or who came to know him through his writings and then by correspondence. His correspondents included men like Boerhaave and Dillenius, older men to whom he dedicated books as a way of establishing his position, and younger men like Alexander Garden from South Carolina for whom correspondence with Linnaeus amounted to recognition of his position among serious naturalists. When the Royal Academy of Sciences was founded, Linnaeus could arrange membership of it for those of his circle whom he most esteemed.

Garden, like many men of science of the time was a doctor, who devoted his spare time to science, and hoped to do more of it as he got older and passed some or all of his duties onto a junior partner – a scheme disturbed in his case by the death of his younger brother, and then by American Independence. Such men remained very important in the sciences down at least to the middle of the nineteenth century. Another group of Linnaeus' correspondents were wealthy merchants like Peter Collinson and John Ellis, for whom, again, science was a hobby, and who might have a splendid garden (like Clifford, Linnaeus' patron in Holland) or who might, like Ellis, do fundamental research themselves. The scientific community was a loosely-knit one, with some full-timers and a great number of others devoting some time, attention, and money to the work. The same was true, in Linnaeus' day, of chemistry: Berthollet was a doctor, Lavoisier a civil servant, Priestley a Unitarian minister, and Cavendish an aristocrat; but in natural history the pattern was to last a good deal longer, and indeed with local natural history societies making surveys of birds and flowers, it still survives. Linnaeus' ideas of classification and naming, and the interest in plant distribution that followed from them, opened up new directions in which the amateur as well as the research student could do useful work.

Linnaeus' trivial names often are geographical – *africanus* or *americanus*, for example – and reflect the information he had received from a student or a correspondent, who had sent him a specimen, a picture, or a description. In the descriptions of animals he also put a brief note of where it was to be found. The epithets can be misleading, and the valid names of some Indian plants include the specific term *chinensis*, for example. One of his correspondents was J.C. Mutis, of Santa Fé de Bogotá just north of the Equator in what is now Colombia; here Linnaeus read his atlas wrong, and supposed that his specimens were coming from Santa Fé in New Mexico and therefore attributed to that arid region plants which are in fact found in the high Andes. From 1760 to 1816 Mutis collected and described the plants of the region then called New Granada, under the auspices of the Royal Botanical Expedition sponsored by the King of Spain. Mutis trained Indians to paint splendid pictures of plants; but some little local difficulties, like the Revolutionary and Napoleonic wars, and then the struggles for independence of the South American colonies, put an end to the idea of publishing the results. It was not until 1954 that the first volume of what will clearly be a magnificent flora appeared, after what must be a record for delayed publication. It is jointly sponsored by the governments of Spain and Colombia, and which will occupy some fifty large folio volumes. This expedition was not directly prompted by Linnaeus; but as Mutis told him in a letter, 'our illustrious Viceroy, just arrived in this town from Spain, is a most ardent promoter of science... He generally enters into conversation with me, after dinner, about you'. It was Linnean science that the Royal Expedition was engaged in; and although the work was so slowly published, specimens from Mutis had got to Linnaeus so that many of the plants were published by him, with due attribution.

Of all the sciences, botany is perhaps that which seems to be most associated with women, and there have been and still are many very distinguished women illustrators of plants. In systematic botany, this was not the case in Linnaeus' day; there were women patrons, like the Princess of Wales, Augusta, mother of George III and the founder of Kew Gardens, and the Duchesses of Portland and of Norfolk; but when Jane, the daughter of Cadwallader Colden of New York made 'profession of the Linnean system', 'perhaps the only lady' that did so, Collinson reported the phenomenon to Linnaeus in a letter of 1757 as something of which he might be proud. She had classified and drawn a number of plants from New York State; and Linnaeus daily named a plant after her. In the eighteenth century, the real science was mostly done by men. This was no doubt for reasons imposed by society, for Eve, with her urge for knowledge to be like God, and Pandora, with her curiosity, leaving only a vestige of hope behind, seem like the foremothers of modern science. One would not expect women to be behind in pursuing a science dealing with both the beauty and the order in the world; and by the nineteenth cenury they were duly there, though usually still concerned with the less controversial parts of it.

While Mutis was really on the edge of the civilised world and remote from any centres of science, Linnaeus at Uppsala was himself on the fringe of the international community. He came to Holland originally bringing his manuscripts with him, and thus with his ideas already formed in a Sweden that was at that time a rather provincial place. He had not had the benefit of the large collections he would meet in Holland, and also in Paris, nor had he met any of the leading figures in the relatively small world of natural historians. In this he was not alone in his period. Benjamin Franklin in Philadelphia worked out in provincial isolation his theory of electricity; and Dalton was later cut off in Manchester from the sophisticated scientific world with its centres in London and Paris. It is sometimes an advantage not to be too sophisticated; all these three produced science that was simpler than might have been expected in a metropolitan centre, where natural classifications, problems of forces between particles, and discussions of chemical affinity were preoccupying the experts. The ideas of all three made headway when accepted and discussed in major scientific societies and academies, but perhaps could not have originated there. In the science of the twentieth century, stress is laid upon the work of teams, but this may be just democratic rhetoric; in earlier centuries, natural history and exploration necessitated some kind of team (like the crew of *HMS Beagle*), but the ideas – such as those of Linnaeus or of Darwin – were worked out by one person, at first in relative isolation and then in discussion with others.

A background such as that of Linnaeus or Dalton can produce both strengths (like tenacity) and weaknesses (like not really seeing what your contemporaries are getting at). Linnaeus never really understood the work of Ellis on the corallines; this was, for the time, quite technical, and so the lapse is perhaps not surprising though it indicates Linnaeus' reliance on morphology. More surprising was his reluctance to accept the view that swallows migrate during the European winter to warmer latitudes. Linnaeus clung to the idea, which he shared with the sagacious Gilbert White, that the swallows gathered near water, into which they then flung themselves conglobulated into a mass (or perhaps singly) and hibernated at the bottom like frogs. Collinson urged upon him the migration theory, quoting evidence from travellers and sea-captains; but to Linnaeus these were only probable stories, and some other reputable witnesses had seen swallows apparently plunging into water, and had sometimes met somebody who had seen one fished out in a dormant state. Collinson added a physiological argument; that if swallows were adapted to spending the winter in this unusual way for birds, then they would have some anatomical structure that would enable them to carry on life in a hibernating state under water. If no such structure could be found, then the thing must be impossible. Linnaeus never saw the force of this argument at all.

Linnaeus' works abound in invocations to God, and there can be little doubt that he saw God as the guarantor that there was an order of things. This was not a new idea, for the view that the world was a cosmos, an ordered whole made by a wise creator, is to be found in Plato; and in the writings of John Ray, the botanist associated with the 'Cambridge Platonists' group of philosophers, the wisdom of God was manifest in the works of creation. The God of this kind of natural theology could become the mere First Cause or celestial clockmaker, the God of the Deists who was no more than a benevolent creator, who made the world a long time ago and left it to run itself. Such a God could be fully understood; His ways were our ways, and He could be congratulated upon a nice piece of design in some animal or plant, especially if it showed some kind of forsight. Ray realised that he had not really found out the plan, the natural system, on which the animal and vegetable kingdoms were founded; and so did Linnaeus. Ray's systems were a mixture of the natural and the artificial, the ideal and the convenient; Linnaeus was perhaps more logical, in proposing to separate convenience from truth to nature. His sexual system was confessedly artificial; and he agreed that to form a natural system would be a splendid long-term objective. Such a system might one day replace his convenient artificial system, perhaps. With his belief in a truly ordered world created by God, Linnaeus was no nominalist simply giving arbitary names to things for economy of thought. He was sure that there was an order, which would in time be found out.

His disciples in Britain took the same line, and the Linnean system prevailed in the English-speaking world down to the

1820s. James Smith, who had bought the Linnean collections, edited in 1821 a valuable collection of letters of Linnaeus and other naturalists. One of Linnaeus' correspondents remarked that a system like the sexual one is 'intended to furnish the learner with unexceptionable characters', that is to exclude borderline cases, whereas in the natural method 'in studying affinities, we seek out the hidden chain of nature'. Smith glossed this passage, remarking that Linnaeus 'better understood' this point, 'when he asserted the necessity of an artificial system for practical use, and of the study of natural orders for a philosophical knowledge of plants. The great fault of the French school is the confounding these two distinct objects'. This pragmatic separation of a practical system and a truly scientific or philosophical study meant that botanists in Britain found themselves in the apparently odd position of classifying and naming plants according to a system in which they did not believe. This did not worry many of them; on the contrary, they believed that those who were carried away by 'the spirit of system' in other countries had made the mistake of prematurely looking for truth rather than consistency and coherence, and the further mistake of thinking that they had found it. For opposed to the clear conventions of the Linnean system were a range of supposedly natural systems; and it was only when these began to cohere that a new generation of English-speaking botanists gave up Linnean orthodoxy.

The Americans, with great numbers of expert botanists, took especially readily to the Linnean system. In England its prominent supporters included not only Smith but also Banks, who became President of the Royal Society in the year, 1778, in which Linnaeus died. In 1768–71 Banks had sailed with Cook in *HMS Endeavour* to Tahiti and then on to New Zealand and to New South Wales. He had paid his own way as a natural historian, and brought with him a team of artists and also one of Linnaeus' pupils, Solander, who had come to England and had had various jobs there. In New Zealand, and even more on the eastern shores of Australia which were previously unknown to Europeans, they found a rich harvest of plants that had never been described or named by botanists. Botany Bay got its name from their labours there; and Banks became the great promoter of English colonisation in Australia. Here again the efficiency of the Linnean system for classifying and naming plants, even of completely unfamiliar kinds, recommended it to Banks and Solander.

In France, as the note by Smith suggested, the Linnean system had never made the same headway. Linnaeus had corresponded with Antoine and Bernard de Jussieu, who were working towards a natural system for plants; but his system had never gained the approval of Adanson, who had sought a wholly-empirical natural method in which all characters were taken into account and none of them was given any extra weighting. The younger Antoine de Jussieu, son of Bernard and nephew of the Antoine who wrote to Linnaeus, published in 1789 his *Genera Plantarum* on the natural method, and therefore proposing quite different genera from those in Linnaeus' book of the same title. Meanwhile, the great figure in zoology had been George Leclerc, Comte de Buffon, who was in charge of the Jardin du Roi in Paris. Between 1749 and 1804, the last volumes being posthumous, he produced his *Histoire naturelle* in fourty-four volumes; and this magnificent compendium became one of the great books of the eighteenth century. It outraged the Americans because Buffon wrote that animals there degenerated by comparison with those of Europe and Asia, and that no really large animals were to be found there. It also brought upon him the wrath of the theologians of the Sorbonne because of some of his speculations, notably that some species might have changed by degeneration since the creation. Buffon tried to define animal species more naturally than Linnaeus, in terms of interfertility rather than of morphology. For Buffon, a species was something like a breeding population. Where Linnaeus' works were generally systematic and rather dry, Buffon's was an often shapeless, but entertaining, compilation.

Where Buffon's criterion might have saved a lot of trouble was with man, for sailors had long since proved that all the various races of man were interfertile and therefore cannot be different species. If some men belong to a different species from the dominant race, then that race could enslave them just as we enslave dogs or horses, without moral censure; and with the revulsion against the slave trade at the end of the eighteenth century this argument was used to reinforce the Christian doctrine of the brotherhood of man. In the first and the tenth editions of the *Systema Naturae*, Linnaeus had placed man in a group with the monkeys. In the first edition, *Homo* was a separate genus with only one species but four varieties; by the tenth, the canonical one, where the binomial names were in use, under

Homo sapiens we find the varieties 'wild' (wolf-children), 'American', 'European', 'Asiatic', 'African', and 'Monstrous'; but within the genus there is another species, *H. troglodytes*, the orang-outang. The pious Swede thus put an anthropoid ape (the only one he knew about) in the same genus with mankind, much closer than the Darwinians were to do – the two sorts of rhinoceros that he knew similarly formed one genus.

It was indeed widely accepted at the end of the eighteenth century that man differed from the orang-outang mainly in the power of speech. There were those who went so far as to believe that society had corrupted man, whereas his congener the ape was innocent and benevolent; this idea is found in Lord Monboddo's work on *Ancient Metaphysics*, 1779–99 (Monboddo being a Scottish judge), and is made fun of in T.L. Peacock's novel *Melincourt*, 1817, in which the hero is an orang-outang, Sir Oran Haut-ton, who becomes a Member of Parliament and a baronet. If the savage could be deemed noble, then how much nobler must be the really wild member of our genus.

At the end of the eighteenth century, then, the idea that man was very close indeed to the apes was something to be earnestly proposed or to be satirised, but it was not so shocking as to be unmentionable or to provoke denunciations from pulpits. This was not even because nobody made any evolutionary speculations about actual relationship; though a Frenchman in Diderot's circle, Delisle, wrote that apes were degenerated men. The belief in a Great Chain of Being was still strong enough for it to be taken for granted that there were no jumps in nature, and that there would be creatures between man and the monkeys – the orang-outang was simply this missing link. The confidence of mankind was sufficient for nobody to feel threatened by apes for cousins as people were a century later. Linnaeus' system was dry and artificial, as Copernicus' writing had been mathematical as hypothetical also. One of the factors that made discussion of man's place dangerous was the revulsion against the French Revolution. The materialism of French science, in which man was no more than an animal or perhaps a machine, seemed to have led to social disruption culminating in the Reign of Terror and leading on to the wars of conquest under Napoleon. In the same period, France was the scientific centre of the world; and men of science in other countries looked with mingled awe and horror at what was going on in the enemy capital, Paris. It was there that the natural methods of classification were being introduced into botany and zoology to replace those of Linnaeus, to the distaste of James Smith; and it is to these methods that we must now turn.

4

The Shape of Nature

P<small>ARTICULARLY</small> in its first form, as a series of tables, the Linnean system had a shape; animals, plants or minerals were being placed in a kind of grid, and one could appreciate the pattern. Such a tabular classification is perhaps the most satisfying kind of ordering, because the order is apparent to the eye as well as to the mind. For plants, the sexual system had the further advantage of being quantitative (in an elementary way), and of applying the same criteria throughout the system; these two features meant that there should have been no borderline or indeterminate cases. Even in Linnaeus' lifetime it was clear that there were some; the numbers of stamens and pistils in a species are very constant, but there are, for example, some species of willows in which the number of stamens varies. This was pointed out to Linnaeus by Albrecht Haller of Göttingen, with whom he had a tense relationship, both sides admiring the work of the other but Haller being always unsatisfied with an artificial arrangement. Had there turned out to be many such exceptions, the sexual system would not have been acceptable at all; but even some exceptions weaken a system which was designedly artificial, because they make it less certain and convenient.

For the animal kingdom, Linnaeus could not devise so neat and quantitative an arrangement; within the mammals, the teeth provided him with something that could be counted and used in classification, but apart from his binomial names, his system was not especially convenient in leading to the rapid placing of animals in classes, or in avoiding ambiguities. The same was true of his mineral classification. No overall plan was clearly discernible in these kingdoms, and Linnaeus' system recommended itself because it was consistent and rule-governed rather than because it threw very much light on the order of nature.

Most of Linnaeus' contemporaries and successors, while using his system (and modifying it as they discovered new species and genera), believed in a different kind of order: the Great Chain of Being. This was an order of things that could readily be imagined or depicted, at least in a general way, for everything lay on a linear series. This vertical chain, ladder, or scale stretched up from unorganised matter through crystals into plants, on to animals, and then up to man, with perhaps angels above him. It had some basis in observation – we can still talk intelligibly, it unwisely, about organisms being higher or lower in the scale of nature – but its central idea was theological. It had the advantage of not merely showing how things were arranged, but also why they must be so. God did not have to create anything, being complete in Himself; but if He did decide to create a world, then He would make it as full and varied as possible. A creation with void holes in it would be incomplete, like an unfinished painting, and unworthy of God; and voids among the animals and plants, where species could have been created to link existing creatures, were just as bad as spaces devoid of matter which could have been filled up. This principle of plenitude was a guarantee that nature made no jumps. Radical discontinuities were only apparent, and indicated that missing links were still to be discovered. As it was put in verse by Stillingfleet:

> Each shell, each crawling insect holds a rank
> Important in the plan of Him, who framed
> This scale of beings; holds a rank, which lost
> Would break the chain, and leave behind a gap
> Which nature's self would rue.

In the eighteenth century, work such as Ellis' on the corallines helped to fill in the gap between plants and animals; while the orang-outang filled that between monkeys and men. The chain of being provided therefore a kind of 'paradigm', a wide-ranging theory which made certain questions seem especially worth pursuing. But there were two major weaknesses in the theory. The first of these was that a chain or ladder has links or rungs; that is, it is not continuous like a rope. The various species form the links or rungs, but there is nothing in nature to determine the distances between them. If nature really made no jumps, then the image of a chain is perhaps misleading, and every species should shade imperceptibly into its neighbours – but perhaps this does fit the image, for the links of a chain do interlock. In nature, this is not what one finds; it is possible to fix upon species, which generally do have determinate boundaries. Horses come in many shapes and sizes, but they are distinct from donkeys. The two species are clearly close to one another; the question is whether they are close enough to count as neighbour species, or whether we should look for a missing link even between them. In the Great Chain theory, we can never know (unless we found all species merging, which would make naming and classifying very difficult) whether we have got gaps left or not.

Among the larger mammals things are difficult enough, but when we look at the invertebrates, where fresh species were constantly being discovered in the eighteenth and nineteenth centuries, it becomes more and more difficult to arrange them in a serial order moving up from amoebas towards man, and to decide where the biggest gaps are left. To put the plant kingdom below the animals seems intuitively all right, but when one thinks about it it does not any longer seem obvious that a tree or an orchid, displaying perhaps all sorts of curious structures and contrivances, should be placed below a single-celled animal just because it is immobile. Ellis' corallines were on the boundary between plants and animals because earlier naturalists had put them on the plant side of the divide, and he put them on the animal side; but even though he and his contemporaries called such creatures 'zoophytes', which means animal-plants, they really assumed that they would fall on one side or other of the divide. Their placing was not seen as a matter of convention; a real frontier was perceived, and they were on one side or the other, and were not really a missing link between the kingdoms – kingdoms did not shade into one another, in Europe or in Nature, in the eighteenth century.

The second major difficulty about the theory of a chain or ladder was that it assumed that God, in filling the world with His creation, would have made the best of all possible worlds – to think otherwise was to cast doubt on His benevolence, wisdom, or power. Such a world would have been a timeless steady state, for if it were to change it could (since it is the best of all possible worlds) only get worse. The Deistic geological system of James Hutton, who saw in 1785 endless cyclic processes with 'no vestige of a beginning, – no prospect of an end', fitted well with this doctrine, called Optimism; since such a philosophical optimist believed that nothing could really change, and that nothing could really be done about anything, his beliefs were very different from those of an optimist in the popular sense. By the end of the eighteenth century, there were many who were not prepared to see famine and social injustice as features of the best of all possible worlds, who thought that things could be improved, and who rejected this steady state theory in favour of belief in progress, or at least of change through history. Within natural history, the doctrine implied that no creatures could have become extinct – except possibly, like the dodo, through human agency; and this was generally believed down to about the end of Linnaeus' life.

By the late eighteenth century, fossils had been discovered in sufficient numbers to make it clear that they were the remains of creatures, and of creatures different from those with which the naturalists of Europe were familiar. The majority of fossils are shells, and it was in Italy, where the deposits are recent in geological terms, that the fossil shells were found to resemble living forms so closely that their status could not be doubted. In the older rocks of England, enormous ammonites and belemnites unlike anything anybody had ever seen alive, and in which the shells had been often distorted and transformed into quartz, were much more baffling to the taxonomist. By analogy with the Italian shells, their status as organic remains of a former world (to use the title of a famous book of 1804–11) could not be doubted, and fossils had to go from the mineral realm to the animal or vegetable kingdoms. Anomalous objects remained (and still do) where it was not clear whether it was a real fossil or a concretion; and some things that had been supposed plants, like the 'stone lilies' of Derbyshire, turned out to be animal, the remains of stalked starfish.

In the late eighteenth century, under the spell of the Great Chain of Being, naturalists expected that even if some species had disappeared from Europe, it would be found somewhere else – the wolf, after all, had been exterminated in Britain but continued to flourish elsewhere. In Siberia, in the interior of North America or of Africa, or perhaps in the hypothetical southern continent of enormous size to balance Europe and Asia, *Terra Australis Incognita*, there would be found living the creatures that had been found fossil in Europe. God would not have allowed any species to become extinct in this best of all possible worlds. It is rare to find the complete skeleton of a fossil vertebrate; teeth and odd bones, often broken, are most often found, and to determine species from them was a problem not really solved until the end of the eighteenth century. Thus when Jefferson found the bones of a large creature, he attributed them to a large carnivore, bigger and better than a lion, that had roamed America, and perhaps still did somewhere. Later, when mastodon bones were discovered in Ohio, travellers to the wild west had elephants to keep a look out for as well.

Jefferson's *Megalonyx* was redescribed as a giant sloth by a Frenchman, Georges Cuvier, who was to become one of the giants of zoology. His researches into comparative anatomy gave his great work, *The Animal Kingdom*, an authority that Buffon's compilation had never had. He was especially interested in ichthyology, but fish were less exciting than fossils and in the public eye he was famous for his work on reconstructing extinct creatures, and for his classification of animals. The Royal menagerie and botanical gardens had been reorganised at the Revolution into the Jardin des Plantes in Paris, where the Museum of Natural History, the gardens, and the zoo still go on. The Museum, to which professors of the various branches of natural history were appointed, was the only scientific institution to survive unscathed all the ups and downs of the revolutionary years; and like the other revived or new scientific institutions, it flourished greatly in the years of Napoleon's First Empire. Cuvier was one of the professors, and soon became the most prominent; he later also became Permanent Secretary of the Academy of Sciences and had the task of delivering polished *éloges* on deceased colleagues.

Under Napoleon, Paris was being rebuilt as an imperial capital, and in the quarries of the Paris basin fossil remains were found in considerable quantities. Particularly striking were the fossils of large mammals, and these Cuvier (whose professorship was in vertebrate zoology) took over as his particular province. The problem was that there were very few complete skeletons, and that the bones and teeth of various kinds of creatures were found all jumbled together. The problem facing the zoologist in this valley of dry bones was to know like Ezekiel how to prophesy and make them live. Cuvier took from Aristotle the principle that all the parts of creatures are made to work together, and with this teleological principle of correlation (as he called it) he was able to make the dry bones respond to his word and come together into the various creatures that they had composed. Predators must have powerful jaws and claws and can have relatively simple digestive systems; if one finds a jaw with powerful canine teeth, then it cannot have been associated with long, thin cloven-footed legs, because these are characteristic of herbivores which need flat grinding teeth. The principle seems simple and indeed obvious; but Cuvier's judgement became so good that it was said that he could identify an animal from a single bone. Some bones are better than others for this kind of thing; but the feat was performed by Richard Owen, an Englishman of Cuvier's school, who identified the broken bone shown him by a sailor from New Zealand as coming from a large ostrich-like bird, and was proved right when the skeleton of a moa was found not long afterwards. Owen formally described the fossils Darwin collected on the voyage of *HMS Beagle*.

Cuvier had the resources of the Museum at his disposal, and could compare the unknown bone or tooth with the specimens there; and Owen similarly had the collections at the Royal College of Surgeons, and later at the British Museum. Such great collections were essential for anybody who wished to build up the knowledge of comparative anatomy necessary for determining species in this way; and with the museums grew up a generation of closet naturalists, of whom it was said that some did not know that the animals they looked after had not always been stuffed with straw. Defining creatures from their bones, or at least from their corpses preserved in spirit, meant that one did not trouble too much about how they behaved, apart from such general features as were required for applying the principle of correlation.

Cuvier did not merely separate the sabre-toothed tiger, the cave bear, and the mammoth out of the jumbled bones of the Paris basin; he discriminated various species, for example of woolly rhinoceros. He could indeed reconstruct a complete mammalian fauna, very different from that now found in France. By the opening years of the nineteenth century, scientific navigators like Cook, Bougainville and la Pérouse had exploded the enormous Southern Continent, but had begun the exploration of Australia: which turned out not to have mammoths roaming about in it, though it did have creatures to daunt the taxonomist, like the duck-billed platypus. Jefferson bought the vast territory called Louisiana from Napoleon's government, and sent Lewis and Clark to explore it; and again, while they had some adventures with grizzly bears, they encountered no mastodons or giant sloths. Under the scientific direction of P.S.Pallas, various expeditions had crossed Siberia, where they had come across edible mammoth corpses deep frozen near the mouths of rivers, but had never seen a living one. Even Africa was at last being explored by men with scientific knowledge like Mungo Park, and there, too, there was no sign of the kind of large mammals of which Cuvier had found the fossil remains. One might miss noticing a shellfish or two in remote waters, but one could hardly fail to observe mastodons or sabre-toothed tigers had there been any; and thus the doctrine that had supported the Great Chain of Being, that this was the best of all possible worlds and that its fauna and flora must be unchanging, had to be given up. The fossils of the Paris basin represented an extinct mammalian fauna.

Cuvier invoked a series of catastrophes to account for extinct faunas, for different fossils were soon found to be characteristic of different strata, indicating a series of extinctions; the frozen mammoths must have been caught in a dramatic change of climate (if it had been slow they would not have been quick-frozen and still edible), and other creatures must have perished wholesale in floods of terrifying proportions. In geology, the tension between those who like Cuvier invoked catastrophes to explain great changes and those who like Charles Lyell believed that everything was to be attributed to ordinary causes operating over immense tracts of time, went on through the nineteenth century, with both sides having to make concessions. Geology became an historical science, and in history neither the 'steady state' nor the 'radical innovation' schemes really fit everything; there are real and explicable changes, sometimes slow like the Industrial Revolution and sometimes rapid like the French Revolution, and yet much goes on unchanged through them. So it is in geology, though large-scale catastrophes quite as sudden as Cuvier's are no longer admitted.

Cuvier also rejected the Great Chain of Being because he did not believe that all animals could be placed on one line of increasing complexity. Instead he invoked four great branches, making his classification a tree rather than a ladder. His tree had no evolutionary implications, though this is perhaps difficult for us to appreciate, for Cuvier was convinced that species were fixed. The four branches were the vertebrates, the molluscs, the articulata (which included the insects), and the radiata (like the starfish and the sea-urchins, having a radial symmetry – in fact, a holdall group.). It was possible to place creatures as higher or lower on their particular branch, but there was no point in comparing animals on differett branches to determine which was higher. Creatures were classified according to multiple criteria, weighted so that for vetebrates, for example, it was the teeth and bones which were of paramount importance and which meant that extinct and living creatures could be compared. Later authors sometimes disagreed about how many great branches there should be, and about which branch anomalous creatures like the barnacles should be put on to; but Cuvier had given the early nineteenth century a pattern which could be visualised, and yet seemed to represent a natural system.

What is perhaps curious to us about Cuvier's *Animal Kingdom* is that like Linnaeus he worked downwards, starting with man; down to Darwin, man remained the measure of all things in zoology and one went from the most perfect species down to simpler ones. The book first appeared in four volumes in 1816; after Cuvier's death, his 'disciples' produced a splendid edition in seventeen volumes between 1836 and 1849, and an English version appeared in fifteen volumes with contributions by several eminent zoologists between 1827 and 1834. For Darwin, Linnaeus and Cuvier were his two gods, until he read some Aristotle in translation and put him in the pantheon too. Cuvier did not carry all before him, for in 1822 John Fleming published his *Philosophy of Zoology* in Edinburgh. In the British tradition, he was an amateur – there were few who could be professionals like Cuvier and his French contemporaries – being a clergyman, knowledgeable in zoology, whose work was taken note of although it never gained a wide following. He proposed an artificial system, bifurcating so that one started with divisions that split the group into approximately equal parts. As a rule, orders should have two genera, and genera should have only two species; though the *Bimana* order had only one genus with one species, namely man, who was thus much further from the apes than Linnaeus had put him. Fleming was like Linnaeus in that he recognised the desirability of a natural system, but felt that the convenience of an artificial one would be such as to make it generally prevail.

Fleming was a firm vitalist, believing with the great physiologist John Hunter that some vital principle was required to account for the phenomena of life, and that organic matter only became fully subject to the ordinary laws of physics and chemistry at death. So eminent a chemist as Sir Humphry Davy, President of the Royal

Society and, with Raffles, the founder of the London Zoo, would have agreed with him; vitalism was not eccentric in Britain at that time, and the vital principle was what distinguished the animal and vegetable kingdoms from the mineral. Fleming used the term 'Theory of Evolution', but not to mean what we would expect: he meant the idea of preformation, that the embryo existed fully formed but minute in the spermatazoa or the ovum, and that fertilization simply triggered off its 'evolution' into a baby big enough to be born. He regarded this theory as demolished by the birth of hybrids – there should be no mules, but only horses or donkeys if preformation were correct. This use of 'evolution' is one which remained current through the first half of the nineteenth century, which is why Darwin and his allies avoided using the word for their theory as far as possible. The term also carried with it overtones of orderly, progressive growth which were alien to the Darwinian world-view.

Fleming did refer to what we might call 'evolution' too, when he wrote of about the geological succession of organisms, and rejected the idea that 'descent with modification' might have changed former species into those with which we are familiar. There was, he thought, no evidence for this hypothesis, and no missing links had been turned up: he preferred to suppose that different creatures had been called into being at different periods, as the state of the Earth had become suitable for them. This view seems pious but unhelpful, because it goes no way towards explaining why some species have had much longer histories than others, some 'living fossils' holding their own through enormous changes in the rest of the fauna and flora. Fleming recognised a 'wasteful war every where raging in the animal kingdom', but he thought that Providence had assigned limits to it; and urged zoologists to 'examine the existing causes of change in the animal kingdom, in order to comprehend the alterations which have already taken place, or to anticipate those which may yet be produced.'

In 1822, a theory of descent with modification based upon existing causes might even have been described as 'Darwinian', after Dr Erasmus Darwin the physician and poet, who had been associated with the engineers Boulton and Watt, the potter Josiah Wedgwood, and the chemist Joseph Priestley in the Lunar Society of Birmingham. This group met at the full moon, when there was light to travel, and discussed scientific matters. Lichfield, where

Darwin lived, was the birth-place of Samuel Johnson, and there a translation of Linnaeus' works on plants had been published, by a society of which Darwin was the moving spirit, between 1783 and 1787. Darwin then wrote up the sexual system in a series of long didactic poems between 1789 and 1803, the titles being *The Botanic Garden, Zoonomia, Phytologia,* and *The Temple of Nature.* These poetic effusions on the loves of the plants were well-received; but Darwin's radical sympathies – especially the favour with which he regarded the French Revolution – lost him friends and readers, especially after he had been satirised in the right-wing periodical, *The Anti-Jacobin,* in a spirited piece on the loves of the triangles. The poems, especially *Zoonomia* and *The Temple of Nature,* contain a progressive view of nature, and expound the view that all animals are descended from one living filament and had diverged in character to meet different circumstances.

By 1802 when Darwin died, the idea of writing science in verse (a tradition that goes back to Lucretius' poem on atomism) was almost extinct; and although his works were clearly very learned, with notes and appendices often exceeding the poems in length, they were not taken seriously by a generation that had come to expect science to appear in more sober form, and to promise (as in Davy's lectures, to which the fashionable world flocked) utility as well as entertainment. Darwin's works could be safely written off as speculative, characteristic of their day, and not to be taken seriously; the best-sellers of the previous generation often thus become unreadable and are unread. Charles Darwin claimed not to have come across his grandfather's views when he was developing his own; and certainly echoes of the older man would not have helped in the task of getting men of science to take a theory of organic development seriously. Down to 1859, speculative theories of an evolutionary kind were well known, but they had few respectable adherents in the scientific community.

The great exception was J.B.Lamarck; and although his evolutionary writing was not widely read, his views were known – though generally through his critics, who included Cuvier and Lyell. Lamarck was one of Cuvier's colleagues at the Museum in Paris, and he had specialised in invertebrates. He was one of the coiners of the word 'biology', which unlike natural history was concerned only with the plant and animal kingdoms; the nature and origin of life preoccupied him. In 1801 he published a work on the classification of invertebrates, and from 1815–22 a major study of their natural history and grouping, in seven volumes. This work his contemporaries could not but respect, for into those regions where Linnaeus had vaguely grouped worms and insects Lamarck brought order; and moreover he tried to establish a natural system (following his early training in botany) for these organisms, which greatly outnumber the vertebrates both in species and in individuals.

Although one might expect the coiner of the word 'biology' to resemble a modern, specialised scientist, Lamarck was not like that; he took an interest in chemistry, and formed a theory of his own in opposition to that of Lavoisier, and also in meterology, preparing between 1800 and 1810 annual reports on the weather. This range of interests was characteristic of men of science of the eighteenth century, and of Lamarck's contemporaries in England; but in Napoleon's Paris such a range, from the wildly speculative to the tediously empirical, was looked at askance, and Lamarck was regarded by his contemporaries as unsound and carried away by the spirit of system. His evolutionary writings confirmed these suspicions.

Lamarck could not, like Cuvier, give up the Great Chain of Being apparently without a pang; and rather than give it up, he sought to preserve it while taking account of the creatures which had disappeared. His conclusion was that such creatures had not been wiped out, but had changed: so that whereas Charles Darwin's theory was closely associated with extinction – the fate of any organism that cannot maintain its adjustment to its surroundings – Lamarck's was based on a denial of extinction, and not merely on the survival of the fittest but on the transformation of the whole species. Lamarck's critics saw in his theory the claim that there was some tendency to improvement inherent in nature, and that plants and invertebrate animals wanted their offpsring to get on in the world and turn into something higher in the scale; but this was a caricature, though based on some unhappy phrases. He moved from a position where he allowed limited change, as Linnaeus and Buffon had done, to limitless change, in the first decade of the century; and he came to refer to nature's plan of increasing complexity, but probably only as a metaphor. Use and disuse were very important in promoting the growth and the shrinkage and ultimate disappearance of organs; the environment did not mould organisms directly, but it did passively determine which way their development went.

It is surprising to find a taxonomist thus urging that what he is classifying is in fact unstable; Cuvier's insistence on the fixity of species seems more understandable, and Lamarck's contemporaries were able to use the geological record, in which all fossil creatures seemed to belong to definite species rather than to lie along some continuum of change from the past to the present, as an argument against his theory. The advantage Lamarck obtained was that he retained an overall pattern, in the form of the Chain of Being; perhaps better seen as a ladder up which creatures were continually moving. At the bottom of the ladder, spontaneous generation was bringing into being creatures like intestinal worms; while at the top we might hope for improvement in man as we evolved away from our apish ancestors As tends to happen with clear overall patterns like this, the single line of development proved impossible even for Lamarck, and by 1815 he had come to propose two separate series for the invertebrates, which was slightly simpler than Cuvier's three great branches devoted to these animals but which destroyed the basis in a chain or ladder.

Lamarck's excursions from taxonomy into chemistry or evolutionary theory came to be seen as a dreadful warning of how 'the spirit of system' characteristic of parts of eighteenth-century science could lead anybody astray – Cuvier was required to tone down passages in his *éloge* of Lamarck where he had been too forthright in his criticisms – and the *Philosophie zoologique* of 1809 in its two substantial volumes was little read. Cuvier satirically remarked of the theory it contained: 'ducks by dint of diving became pikes; pikes by dint of happening upon dry land changed into ducks; hens searching for their food at the water's edge, and striving not to get their thighs wet, succeeded so well in elongating their legs that they became herons or storks. Thus took form by degrees those hundred thousand diverse races, the classification of which so cruelly embarrasses the unfortunate race that habit has changed into naturalists.' The classification of invertebrates by this newly-evolved natural group did however proceed apace; and it was in this field that a new pattern for a natural classification was perceived, and the Quinary System was born.

The two great protagonists of this system were W.S.MacLeay, who invented the system, and William Swainson, who applied it to zoology generally, in a series of works; together they ensured that between about 1820 and 1835 systematists, and indeed

zoologists generally, in Britain had to take their proposals
seriously, even if they did not accept them. Some of MacLeay's
writings were translated into French, but the ideas never caught
on there or elsewhere; and for this period much taxonomy in the
English-speaking world was carried on on lines different from
those accepted elsewhere. The episode therefore tells us
something about national styles and international relations in
science. The sciences are often supposed to be international, but
things are more complex than that and one can find certain
emphases that are characteristic of one nation. In Britain a kind of
ancien regime survived well into the nineteenth century, as it did
not in most European countries; a different social system, and a
different attitude to religion, was to be found on the two sides of
the English Channel. What is curious is that MacLeay and
Swainson were both francophiles, and admirers of Cuvier; they
were not like Fleming working up an artificial system, but
supposed that they had got the key to the natural system and had
found out the pattern of the creation.

MacLeay was the son of an eminent entomologist, who had a
very large collection of insects, had been elected Secretary of the
Linnean Society, and was in 1825 appointed Colonial Secretary for
New South Wales – later he became first Speaker of the Legislative
Council of the colony. His son also had a legal and diplomatic
career; in 1839, after being President of his Section of the British
Association, he too went to Australia where in a more genial
climate he devoted himself to the enlargement of his father's
entomological collection, to the encouragement of the young
T.H.Huxley voyaging on *HMS Rattlesnake*, and to the
development of a splendid garden. He had gone up to Cambridge,
taking his degree in 1814, and had in 1815 been appointed a
member of a commission sent to France to clear up claims that had
arisen from the wars; there he met Cuvier, Geoffroy St. Hilaire,
and Latreille. On his return he published, in 1819–21, his *Horae
Entomologicae*, which was the first account of his system. Darwin's
work on barnacles was undertaken partly to refute this famous
but exceedingly rare book; curiously enough, it was MacLeay who
persuaded Darwin to undertake publication of the *Zoology of the
Voyage of HMS Beagle*, which appeared with government support
in 1838–43 with descriptions by various experts.

MacLeay's arrangement of animals was called the circular or
quinary system. It was based upon the idea that the various

groups of animals could not be fitted into linear sequences, with the different members being placed higher and lower. Rather, the groups formed circles, in which the extremes met (as in Coleridge's philosophy). If, in MacLeay's view, the naturalist took the various characters of an organism into account, he would be able to see this circular arrangement; and he would find moreover that this circle touched other circles so as to form a bigger circle; thus a genus forms one circle, which forms part of the circle of an order, and so on. In place of the ladder or chain, or of Cuvier's tree, the system is one of wheels within wheels (if we may borrow another metaphor from Ezekiel) like the logical diagrams of Linnaeus' contemporary, the eminent mathematician I. Euler. The term 'quinary' was attached to the system because the various circles were seen as having five members, and the system thus arranged things in fives.

Each set of circles consisted of two big ones, containing the highest or 'typical' group, and the next or 'sub-typical' one; while below them was another equal circle, containing three smaller circles occupied by the 'aberrant' groups. In a draft classification of the mammalia of about 1829, MacLeay put mankind into a set of circles, treating the races of man almost as though they were species (which he knew they were not). The top circles were occupied by the civilised groups, the Caucasian and the Mongolian, while the aberrant circles held the Americans, the Negroes, and the Malay/Polynesians, all being savages. He only knew of two tailless apes, which occupied the top circles of the pithecoid arrangement, the tailed apes being the aberrant ones. The system is thus based on the numbers two and three as well as five.

Compared to Cuvier's branching tree, this seems a closed system; and despite its advantage of not putting animals 'higher' and 'lower', one might expect that it would have been difficult to fit in newly-discovered creatures. This did not turn out to be the case, as MacLeay demonstrated when he classified the insects that Thomas Horsfield, a protégé of Raffles in Java, collected; these he published as *Annulosa Javanica* in 1825. Later critics have seen the quinary system as thoroughly artificial, and indeed by about 1840 it was thought of rather like Lamarck's evolutionary ideas, as an awful warning and an example of how not to proceed. But in 1819 the natural method had not yet, in England, displaced the Linnean, and it was some years before the classification of Jussieu

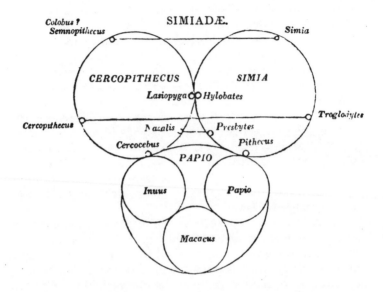

The monkeys arranged according to the Quinary System, from W.Swainson, *The Natural History and Classification of Quadrupeds*, 1845 ed., p.72.

was accepted in botany although in 1817 Banks had begun to waver in his old age, writing to Smith:

> I fear you will differ from me in opinion, when I fancy Jussieu's natural orders to be superior to those of Linnaeus. I do not however mean to allege that he has even an equal degree of merit in having compiled them. He has taken all Linnaeus had done as his own; and having thus possessed himself of an elegant and substantial fabrick, has done much towards increasing its beauty, but far less towards any improvement in its stability.

To see a system of classification as a country house undergoing improvements is perhaps unusual, and characteristic of Regency England (though science as a building was a Baconian idea), but it may remind us that one need not perceive major changes as revolutionary, and that natural methods were novel in England. Even the magic number five, which makes the system look artificial to us, was found often enough in nature – in fingers, toes, and the arms of starfish – to be a reasonable candidate if there was to be any one number behind zoology. That people in England should expect there to be such a number, or at any rate to expect that taxonomy should have some numerical basis, indicates perhaps the long survival of the Linnean system with its numerical basis for the classification of plants. In MacLeay's case, it is perhaps a feature to be expected in one who came to zoology from mathematics. The prestige of mathematics no doubt overcame the scruples of those who suspected that no sytem which keeps dividing into equal numbers, whether twos threes or fives, can really reflect nature.

MacLeay's most important convert was William Swainson, one of the most talented painters of natural history that there has ever been – his birds are particularly striking. There are few people, and there were less in the early nineteenth century, who can work up much enthusiasm about how insects are to be classified; it is a highly technical matter, and to see experts agitated about it is slightly comic. Swainson took the quinary system out of this recondite sphere, and in a series of works brought it before the public as applied to birds and mammals, and even to man. The system emphasised analogies, but it did not, in MacLeay's hands, require the emphasis on external characters that Swainson gave.

Swainson's father worked in the Custom House at Liverpool and was keen on natural history; at fourteen, Swainson left school to work there too. But he disliked it, and managed to get (through family connections) a post in the commisariat of the army, sailing for Malta in 1807. He spent the next eight years in the Mediterranean region, able to indulge his passion for art and natural history; and on his return after the battle of Waterloo, he chose to retire on half pay and devote himself to the study of nature rather than continue in a public career. In 1816 he set out for Brazil, where he collected plants and animals; becoming, on his return, a Fellow of the Royal Society, on Banks' nomination. As a natural historian, he was not untypical of his generation: 'Bred up with somewhat of aristocratic notions, and accustomed, when on service, to *command* rather than to *obey*, I had a rooted dislike of all commercial affairs, and would rather have gone once more on active duty than have sat behind a desk' – much the same might have been said by his contemporary R.I.Murchison, the geologist. Such men were larger than life, tenacious of their honour, prone to

97

monumental rows, and suited neither by education nor temperament for the relatively humdrum activities that we associate with professional science, and which in Swainson's day was perhaps more characteristic of those natural historians who were also medical men.

In 1822, Swainson hoped for the post of Keeper of Zoology at the British Museum, and collected some impressive testimonials. But Davy, who as President of the Royal Society was one of the Trustees, succeeded in getting it for his friend J.G.Children, with whom he had had an unsuccessful partnership in gunpowder-making. Children was a mineralogist who knew no zoology; but just as nowadays a government minister is meant to be an able man who can soon learn the details of his department, so it was not felt then that an appointee should already be an expert. Sedgwick at Cambridge got the chair of geology in the same way, as a good man who would soon learn up the subject. Ironically, Children did not really want the post, but accepted it because he did not want to seem ungrateful to Davy.

Davy wrote to Swainson in the third person but cordially, and Swainson does not seem to have suspected that it was the President of the Royal Society who had fixed the appointment for Children. His suspicions fastened upon the Archbishop of Canterbury, the Establishment embodied, who was also a Trustee; and his friend T.S.Traill, who later became Professor of Medical Jurisprudence at Edinburgh, took up his cause in a number of vitriolic anonymous attacks on the British Museum in influential reviews. The effect of this was to diminish public confidence in the Museum, so that specimens collected on national expeditions were sometimes consigned to private or provincial collections. There were cases in the nineteenth century of posts arranged for unqualified people by influential but unscientific backers, but this was not really one of them. The Archbishop was no doubt too much of a gentleman to emerge from behind the facade of corporate responsibility, and Davy perhaps not sufficiently a gentleman to do so himself; the episode is ironical because it was one of Davy's ambitions to make the British Museum an active centre of science, like its Parisian equivalent (as indeed it did become later in the century).

The episode served to embitter Swainson, whose father died in 1826 leaving him less than had been expected, and he decided to try to live by his pen. The publisher Longman got him to revise some encyclopedic works, and then commissioned him to do an encyclopedia of zoology. This project was later transformed into the zoological volumes, eleven altogether, of the *Cabinet Cyclopedia*, edited by Dionysius Lardner and completed between 1830 and 1849 in 133 volumes. Lardner was alleged to have been christened Dennis, and to have given himself a more imposing name later. Since this was the name of a Sicilian despot, he was nicknamed 'the tyrant', his tyranny consisting in bullying eminent authors like John Herschel and Walter Scott to write volumes of his cyclopedia, and in pushing Swainson to produce a book every three months. Later, his elopement with the wife of an army officer created one of the sensations of Victorian England.

Faced with his great task, Swainson 'had thus only the choice of following the *Règne Animal*, or of working out, as far as possible, that system I had already adopted in theory, and partly exemplified in detail; namely, the circular arrangement of animals on the principles of their affinities and analogies.' He had not hesitated, and rather than follow Cuvier embarked on arduous study in working out the quinary theory. As well as the volumes of Lardner's cyclopedia, he wrote and illustrated a work on flycatchers for the *Naturalist's Library*, which is a beautiful little set of illustrated works now highly prized, and also the volume on birds for the official report, *Fauna Boreali-Americana*, on the various animals collected by Sir John Franklin's expeditions of the 1820s down the Coppermine and Mackenzie rivers to the Arctic Ocean. These volumes were the first scientific publications to be subsidised by the British government, which provided £1,000 to pay for the illustrations; they were intended to be formal and factual reports, but classifying involves theory and Swainson's was the quinary theory, and his volume contains a brief exposition of his ideas, used in the classification of the birds. An official work therefore came to exemplify this rather eccentric taxonomy.

It was in the *Cabinet Cyclopedia* that he had space to work out the circles fully. In 1828 he had visited Paris for six weeks to study the birds at the Museum, partly because he had fallen out with the officers of the Zoological Society in London; while there he got on well with the professors, and was even given a room belonging to one of them in order to pursue his heterodoxy. The first volume to come out was the *Preliminary Discourse*, which was intended to complement that which Herschel had written in the same series. Herschel's, on 'natural philosophy', became one of the classic works on the philosophy of science of the nineteenth century, and was much admired for example by Charles Darwin; Swainson's, on 'natural history', was essentially an exposition of his quinary system, and never achieved a reputation comparable to its model. Later Swainson himself wrote:

> I verily believe, that, had I expressed my convictions in a more subdued tone, many of those who now differ from me would have adopted these views, – at least in a general way; but I am always so delighted with detecting either a new link of relation, or in bringing an isolated fact to bear upon general principles, that my enthusiasm sometimes overcomes my judgement. I forget, in fact, that no one, unacquainted with the other instances of a similar nature, – all converging to the same point, – can possibly attach the same importance to a *single* instance, that I do myself.

We may be reminded of that other gentleman-naturalist who, like Swainson, after a visit to the Tropics, retired to the country for some years to work out his theory, and brought it before the world in a volume which did not fully present the evidence on which it was based – Darwin; science is a cruel goddess, and it is never completely clear to her votaries, until it is too late, whether they have been prophets or heretics.

We can probably judge the system best if, instead of looking at the *Preliminary Discourse*, we look at one of the other volumes in the set, where the ideas are applied to an actual group of animals; the most appealing is probably *The Natural History and Classification of Quadrupeds* – although using Cuvier's term in the title and the text, Swainson also refers to 'the class mammalia' and included the cetacea among the quadrupeds. The book begins with a general discussion of the principles of classification; pointing out that when two systems are built upon the same facts, one must judge between them 'by ascertaining which is most in harmony with others of a more general and comprehensive nature.' Cuvier's system failed here, because it had four divisions of animals where nature seems to have but three kingdoms, and one might expect the numbers to be the same; and because it did not bring out the circular nature of affinities. Swainson remarked, and few philosophers of science would disagree with him, that 'when . . . a theory in one branch of science is so far isolated that it bears no relation or reference to other branches, closely and vitally connected with it, there must, in the nature of things, be some great defect in its construction.' This is indeed what many great scientists have said in criticising the partial or *ad hoc* theories current in their day, and one cannot but feel that if Swainson had resisted the temptation to play with numbers and had simply urged the circular nature of affinities he might have had greater success in getting his contemporaries to adopt 'these views, – at least in a general way'. T.H.Huxley did indeed toy with circles for the arrangement of invertebrates, but never became a quinarian in the strict sense.

Some of the point about the numbers is revealed in the introductory part of the book where there is a discussion of man. Swainson was very unhappy that his contemporaries, even including MacLeay, had treated man with insufficient dignity, classing him with apes in a way that even the heathen Aristotle had refused to do. Material and immaterial natures were not so closely allied that they could be placed in one zoological circle; it was not mere pride that made earlier naturalists separate man from monkeys, but a true appreciation of his natural affinities and high descent. To put him with monkeys would be like putting tigers and zebras in the same group because they are both striped; it would be to put superficial and secondary characters in place of fundamental ones, and in the case of man his highest and most distinguishing property is reason, which is not possessed by other creatures. Even man's physical structure may only resemble that of animals by analogy and not by real affinity; his going erect upon two limbs marks a real difference.

The circular system allowed Swainson to propose 'other and weightier' arguments against putting men with apes. Every being must have, on this theory, two or perhaps three relations of affinity (quite separate from those of analogy) which connect it to its neighbours in the circle. Even to the orang-outang, man's relation was distant and doubtful, and there was no other affinity: 'if this cannot be made out, if man alone, of all the created beings on this earth, stands thus isolated, his relation to the ourang-outang proves to be one of mere analogy.' A vast gulf in reality separates man from the animals; and indeed the Quadrumana form circles of affinity in which there is no room for man. A being cannot occupy a place in two circles, and man could not be placed both among the unintelligent and the intelligent beings; rather he connects these two spheres, but belongs to the circle of spiritual beings rather than material ones. To put him with apes would be like putting the butterfly with the wingless insects because its caterpillar belongs there. As far as man is concerned, the quinary system provided for Swainson a little more precision for arguments that are fundamentally little changed since the Renaissance.

Placing man among the monkeys was an unpopular thing to do in England in the early nineteenth century, because of the implications that man was no more than a naked ape or society no more than a beehive; and Swainson could quote Archbishop Sumner (a successor of his 'enemy') in support of his view of man. But while the placing of man was important, Swainson was not one to truckle to authority and there is no reason to doubt that he believed what he wrote. The justification for the system was not only that it placed man correctly, but that it gave rise to an intelligible pattern for the animal kingdom. Just as nature was divided into three kingdoms, so the animals fell into three groups, or circles: the mammals, the birds, and the cold-blooded classes. This last one is divided into the reptiles, the amphibia, and the fish; and this pattern of three major circles, the typical group, the sub-typical group, and the aberrant group (divided into three) is the pattern repeated throughout the smaller groups; the magic number five is thus compatible with divisions into threes.

The various circles touch each other where there are closely-analogous species; and across circles one can find parallels, at similar or different levels since the pattern of circles is always the same. Having set out the scheme for the mammals, for example, Swainson sets out the corresponding orders of birds, occupying the same circles in their arrangement. The corresponding groups in various patterns may be said to 'represent' each other; the cetacea represent the swimming birds, the carnivores the rapacious birds, and so on. Terms like 'representative' were widely used in nineteenth-century natural history to mean a variety of things: it might mean something like filling an ecological niche, when the kangaroo represents the antelope in Australia, or it might imply some relationship, as when one species of finch replaces another, but with Swainson it has a definite meaning of occupying a similar position in a different series of circles.

The first family is that of the monkeys, and they are arranged in the same pattern of circles, three big ones touching each other, the third one underneath containing three small circles for the 'aberrant' group. The highest or 'typical' group is *Simia*, including the tailless apes; the next, or 'sub-typical' genus *Cercopithecus*, the 'ape-monkeys', and the lower circle contains (in three small ones) the baboons, the Barbary ape and its congeners, and the genus *Macacus* of old-world monkeys. These groups correspond to the orders of mammals: *Simia* (intelligent) to the *Quadrumana*, *Cercopithecus* (malicious) to the *Ferae*, Papio (little tail) to the *Cetacea*, *Macacus* (hare-lipped) to the *Glires*, and *Inuus* (crested) to the *Ungulata*. Such correspondence takes the breath away, even though Swainson does admit that they are no more than analogies; those who criticise their contemporaries for misplacing man need to be careful about not being carried away by their own system. The circles of *Simia* and *Cercopithecus* touch with the gibbon and the Cochin-China monkey, with its long arms. The American monkeys occupy another set of circles, and so do the lemurs; in the latter group, one of the 'aberrant' circles touches the system of the bats, via the flying lemur, and another the circles of the *Cebidae*, the American monkeys. The bats are the fourth family of the *Quadrumana*, and there should be a fifth, an aquatic group corresponding to the *Cetacea* among the mammals as a whole. Swainson did not believe in mermaids as in the old stories, but wrote that some of the records, by men of probity, as well as the symmetry of the system, entitled him to say: 'That some such animal has really been created, we have not the shadow of a doubt'.

Swainson was rather proud that his system allowed him to predict creatures not yet discovered living or fossil, and considered that gaps in the system were an indication of its truth. Unlike the Great Chain of Being, the quinary system could allow for extinction: 'When we reflect on the enormous destruction which fell upon the antediluvian animals, and more especially upon the large races of quadrupeds, there is nothing surprising that their place becomes vacant in the existing series. Were it not so, there would indeed be strong presumptive evidence that our views of the natural system were radically defective, for then we should have no room to insert those fossil animals which of late years have been brought to light.' This feature, on the other hand, made the system very difficult to falsify; a gap in the system could be put down to the imperfection of the geological record. But this was exactly what Lyell and Darwin did to defend their systems too; and in the 1830s the geological record was indeed imperfect.

In discussing the rhinoceros, Swainson could urge the artificial nature of systems built solely upon the number and form of the teeth as characteristic of genera; for if this were strictly applied the Indian, Sumatran, and African species would all be placed in different genera. In general, he insisted upon multiple characters, because he was proposing what he believed to be the natural system; but he was unfortunate or unwise when he came to deal with the Thylacine, called the Tasmanian Wolf. Because of its teeth, especially its large canines, he ventured 'to remove this singular type from the other carnivorous opossums, with which, in general appearance, it has no similitude, and place it among those with which its outward structure, and its teeth, more intimately correspond.' Though familiar with the distinction between characters that indicate affinity and those that are merely analogous, he went for external features and put the animal among the dogs, adding that in its circle of 'aberrant' forms it is probably the aquatic member, because reports of it were usually from near the ocean, and because it seemed to have a flattened tail, which generally indicated a good swimmer.

The move was not as large, from the marsupials to the cat and dog tribes, as it would be for most taxonomists, for the opossums form one of the small aberrant circles in the order *Ferae*, the other two being the shrews and the seals, while the typical and sub-typical circles are the cats and dogs on the one hand, and the weasels on the other. Swainson knew that he was thus breaking up the order *Marsupiata* of Cuvier, which was an artificial 'assemblage, uniting, as it does, animals of the most opposite natures, and of the most dissimilar organisation, merely from the circumstance of their possessing a marsupial pouch'. The structure of the teeth was after all Cuvier's main criterion, and this grouping violated it; and moreover no circular arrangement could be worked out for the marsupials, which can be easily enough fitted into a number of circles if separated from one another: 'we entertain a confident belief that a more intimate acquaintance with their structure than we yet possess, will, at no very distant period, lead to their complete amalgamation with the general mass, of which they form integral, though at present dislocated, parts.'

Creatures like the Tasmanian Wolf do display most strikingly what we would see as convergence, where two creatures from quite different families that follow the same kind of life come to resemble each other very closely. It is probably necessary in classifying to assign more weight to some characters than to others, and by modern standards Cuvier in putting the marsupials together was doing better than Swainson, who was taking more characters into account but was using external rather than internal features. Putting the Tasmanian Wolf with the ordinary wolves indicates another feature of Swainson's classification, that geographical distribution is not taken seriously as a guide. He did separate the monkeys of the Old and the New World, but chiefly one would suspect, because this enabled him to fit them nicely into circles rather than because of any real concern over distribution.

The abiding impression that we get from Swainson is of his concern with the overall pattern, and his anxiety to see similarities and parallels where most of his contemporaries and successors did not. Like Lamarck, he sought for the complete plan of nature, rather than just for a few natural genera or orders; there were gaps to be filled in, but this part of science at least would not be an endless quest, and his God is the comprehensible First Cause of the Deists rather than the inscrutable and personal God of orthodox Christianity. Giving too much weight to external characters in breaking up the marsupials might merely be judged ill-justified or old-fashioned; but there are places where Swainson goes in for elaborate comparisons between quite distinct groups, being convinced that every species or higher group must have its analogues in other systems of circles in the divine plan. Thus he devotes over seven pages to analogies between the *Glires* (a group corresponding roughly to the rodents) and the wading birds, involving resemblances in the shapes of faces, in their provident and sociable dispositions, and in their uniformity in colour. This kind of reasoning was a feature of the comparisons between the Microcosm and the Macrocosm of Renaissance philosophers; and in Swainson's day such parallels were also prominent in the German school of *Naturphilosophie*, of which Lorenz Oken was the most famous representative in biological science. But Swainson does not refer to German sources, and there is no reason to believe that he owed his ideas to *Naturphilosophie* though he was engaged in the same enterprise of finding out and revealing the complex patterns behind the world, which was constructed according to a harmonious plan.

Outside the English-speaking world, a circular classification including five circles, and looking a little like the quinary system was used in the volume on *Crustacea* of P.F.Siebold's *Fauna*

Japonica, 1833–50. This was the last major work on Japanese natural history done before the country was opened up to foreigners, compiled by a doctor who had spent some time on the little island in Nagasaki harbour where the Dutch maintained a small colony of merchants. But there were few prominent converts, even in Britain, to MacLeay's system, although various patterns of circles were widely used to indicate relationships even well after the publication of the *Origin of Species* in 1859, and John Venn brought out his method of indicating class inclusion by diagrams of circles in 1866. In 1839 Swainson, discouraged by the failure of another attempt to get a post at the British Museum and by the death of his wife, decided to emigrate to New Zealand and farm. An American visiting Australasia in the 1850s heard to his surprise that both MacLeay and Swainson were living there, and imagined that they had been exiled to the Antipodes 'for the great crime of burdening zoology with a false though much laboured theory which has thrown so much confusion into the subject of its classification and philosophical study.'

Such an obituary would have been generally felt appropriate to the system by the next generation of taxonomists. By now usually with a post in a museum, they did not feel the need to justify their science by an appeal to the harmony and order of the world as Swainson had done, but were content to practice something more like 'normal science', working out groups without worrying about the overall pattern. In the footsteps of Cuvier, such men constructed a natural system to which Darwin could give a dynamic interpretation; and it is to the coming of this pattern-free – or as it seemed to practitioners, theory-free – natural method that we should now turn. Although there were a number of natural methods in fact, we do not find such idiosyncratic national styles in this more empirical search; but we do find all sorts of interest in the attempt to discriminate and to work out real affinities, because no search for a natural method can be a matter of mere technicalities.

5

Accordance with Nature

Cuvier, Lamarck, and Swainson had tried in their various ways not merely to describe organisms, but also to explain how they fitted into a pattern. In natural history, as in other sciences, there is a tension between those who are content to describe, in terms more precise than those of ordinary language, and those who seek to penetrate beyond the veil and reveal the underlying plan or mechanism. In Swainson's day, the atomic theory in chemistry polarised physical scientists, with some like Dalton urging that the unprovable assumption of atoms led to a real understanding of chemical reactions, while others like Davy believed that atoms represented no more than definite quantities, and that too much talk of them diverted people from experiment into mere empty theorising. Similarly, in electricity, there was a deep division between those who were content to follow Ampère in applying equations that worked in terms of predicting currents or magnetic effects, and others like Faraday, and the founders of the Cambridge school of physics, Stokes and Maxwell, who believed that a satisfactory theory must tell us about the actual processes of nature. One can have a descriptive or an explanatory ideal in any science, whether it is mathemetical, experimental, or observational. In opting to describe, one is not following some paradigm typical only of natural history; but one is in a sense throwing in the towel.

Descriptive science is not going to change anybody's world view. Changes within it are relatively technical matters, of interest only to experts; in Kuhn's terms, such science is 'normal' with a vengeance. But it can be cumulative, it can go with teamwork, and it can be judged sound and respectable. The risks in science are taken by those who go beyond description and seek for explanation, because they are very likely to find themselves in a blind alley. Thus it was with Lamarck and with Swainson, who were regarded by the ends of their lives as eccentrics, who had done good work on invertebrate classification and on illustration, but had then gone off. A small minority of such innovators are the scientists whose names have become familiar to those outside the world of science. But there are periods in the history of all sciences when the time is not ripe for a synthesis and an explanation, which could not but be highly speculative, when the collection and ordering of facts seems to be the order of the day, and when high but impermanent reputations are to be made; and the first half of the nineteenth century was perhaps such a period in the life sciences.

Swainson made much in his writing on classification of the difference between *affinity* and *analogy*. These formed the bonds which held the vast and complex system together, like those which hold in place the atoms in an enormous molecule. The strongest bonds were those of affinity, which linked together the members of the same circle, and perhaps linked the neighbouring species in two circles that touched each other. Affinity is a term taken over from human families, and indicating blood relationship; in churches there used to be Tables of Kindred and Affinity posted up listing the people one was not allowed to marry, such as one's sister or grandmother. In the eighteenth century, the term had come into chemistry, and elective affinities were forces between different substances; nitric acid has an affinity for silver, in that it combines with it, but not for gold.

In chemistry, the term could never have been other than a metaphor from human families; but in biology one would expect to find more than this, and suppose that terms like affinity (and indeed 'family' and 'genus') were being used to indicate real relationships. But for Swainson, as for Aristotle and numerous naturalists in between them, this was not the case. He firmly rejected Lamarck's theory, and indeed the circles required (like the Chain of Being) that species were stable – they might become extinct and leave a gap, but they could not change one into another. Horses and donkeys did not come next to each other in a circle because they had a common ancestor, but because they displayed similarities like those found among members of a human family.

Analogies for Swainson were what linked the various circles at their different levels together, like the weaker bonds of complex molecules. The resemblances between groups of birds and of mammals that he made much of and that seemed so far-fetched were analogies; their importance in his system was that they made the circles into a coherent if complicated arrangement, and provided the evidence of an overall plan, where the pattern of circles kept repeating itself. Some of Swainson's analogies were implausible, but the recognition of affinities and analogies in nature was not new with him; he just gave an importance to the analogies that most of his predecessors and successors have not done.

Aristotle in his zoological writings had made the distinction:

> Groups that only differ in degree, and in the more or less of an identical element that they possess, are aggregated under a single class; groups whose attributes are not identical but analogous are separated. For instance, bird differs from bird by gradation, or by excess and defect; some birds have long feathers, others short ones, but all are feathered. Bird and Fish are more remote and only agree in having analogous organs; for what in the bird is feather, in the fish is scale. Such analogies can scarcely, however, serve universally as indications for the formation of groups, for almost all animals present analogies in their corresponding parts.

Aristotle had thus said what most people must have felt two thousand years later on reading Swainson; that affinities served for the formation of families, but that one could find analogies almost anywhere.

The problem is to know which relationships or parallels are those of analogy and which of affinity. Evolutionists know in principle that the latter indicate common descent, whereas the former do not; but this does not in a particular case help to decide which is which, since we have no privileged access to family trees. Swainson did not even have this guide open to him in principle; and while the distinction has been agreed to be important since Aristotle, there have never been clear rules on how to apply it. At different periods, different judgements have been made about what are affinities and analogies; Aristotle's example of fish scales and feathers as analogous, performing the similar function of covering a body would still be all right, but chemical tests on feathers and the scales of reptiles show such similarity that these are perhaps related by affinity, though more remote than that between different feathers. Just as there are degrees of affinity in families, and one may expect first cousins to look alike while kissing cousins may look very different, so it is with affinity among animals and plants; some of Swainson's analogies seem better considered as more distant affinities, others as real analogies where different organs perform similar functions – like the wings of insects and of birds – and others as mere resemblances, as between the faces of groups of mammals and of birds.

The term usually used in the twentieth century is not 'affinity' but 'homology', which is curious since we all believe that animals displaying homologies (which has little theory-loading, meaning only 'correspondences') are in fact related, and we could use the word 'affinity' without metaphor. Goethe, who wrote a famous novel about human relationships with the title *Elective Affinities* (thus bringing the metaphor back from chemistry), was one who laid great stress, at the end of the eighteenth century, upon the homologies in nature. He did not believe in evolution, and nor did most of those who followed him in Germany and then elsewhere; the homologies did not indicate relationship, but only the common design behind organisms and processes.

Goethe's work on plants indicated to him how various parts in some simple archetypal plant could be regarded as modified in actual plants to produce such complex structures as the flower. Homologous parts were those which had a common structure, but had probably been modified in different organisms to suit their style of life. In zoology, Goethe was particularly pleased that he succeeded in identifying in the human skull a vestige corresponding to the inter-maxillary bone which is a feature of the skulls of many mammals. It did not in the eyes of Goethe, any more than it had in those of Pierre Belon who had in 1555 depicted the skeletons of a man and a bird side by side to show the homologies, detract from the dignity of man that he should share structures with animals; on the contrary, it showed the unity of plan behind nature, penetrable by human reason.

Goethe showed the continuity between simple plants and those in which some of the leaves had as it were become monsters, and turned into flowers; and in zoology, some of his successors worked out similar schemes whereby vertebrae had become a skull, and also followed with particular care and interest the development of embryos, a process described as 'evolution' and illustrating the transformation of a simple organism into a complex one. The speculative German *Naturphilosophie* of the opening years of the nineteenth century, associated with the philosophy of Schelling, placed great emphasis on the unity of nature and its accessibility to the human reason. The world was the product of the divine reason, and because we shared in this attribute being made in the image of God, we could understand it; the mind rightly used had the key to knowledge.

In the physical sciences, this school urged the unity of all forces, and the primacy of force over mere matter; while in the life sciences the emphasis was on unity and on continuity, nature making no jumps. Some later German scientists, most notably Liebig the chemist, reacted violently against *Naturphilosophie* as delusive, replacing experimental work in a laboratory by speculation or empty verbiage composed in an armchair. But there can be little doubt of its importance at the beginning of the century, as the German universities began to revive. As the sciences were taught there in the philosophy faculty, scientists (and others) willy-nilly learned philosophy of science along with science itself. Oken, the most prominent member of the school working in the life sciences, edited an important journal, *Isis,* and was also the moving spirit behind the annual meetings of the *Naturforscher,* those working in science from all over Germany, which were held in a different city each year and which gave an enormous impetus to science there. These meetings were imitated abroad, in the British Association for the Advancement of Science and similar bodies in different countries; they escaped some of the élitist emphasis associated with academies of sciences, and harnessed the energies of those living in the provinces as the metropolitan scientific societies had never managed to do.

Naturphilosophie had its importance in the next generation too, with the postulation of conservation of energy on the one hand, and with the work on archetypes and homologies by Owen and by Louis Agassiz, who were by the 1850s the most eminent naturalists in Britain and in the USA respectively. Owen was a comparative anatomist, in charge first of the museum of the Royal College of Surgeons and then, from 1856, of the British Museum (Natural History) at the time of its separation from the rest of the British Museum and its move to South Kensington. Prince Albert admired him, and he gave classes to the Royal children; those who had to work with him seem generally to have come to detest him. His biology of ideal types to which living creatures approximately conformed seemed very old-fashioned to those like T.H.Huxley, who were trying to introduce the new physiology of a younger generation of Germans, and also the dynamical theory of Darwin in place of the static species of Owen. Because species represented archetypes, links and gradual transformations between them

111

were ruled out by Owen; development, if it had taken place (and Owen sometimes did believe that it had) must have been jerky, with jumps from one properly organised species to another.

Agassiz wrote in 1857 the first two volumes of his *Contributions to the Natural History of the United States*, which are sumptuous tomes mostly concerned with turtles; but the bulk of the first volume was taken up with an essay on classification, which was in 1859 published separately in England, very shortly before the *Origin of Species*. Agassiz, like Owen, is chiefly remembered as an opponent of Darwin, which he was; but it would be a mistake to see his enormously popular writings, or those of Owen, as simply reactionary. Agassiz was born in 1807 in Switzerland, and studied at Zurich, Heidelberg, and Munich. In Germany he was influenced by Oken and Schelling; and then in 1832 he went to Paris to study with Cuvier, who became his hero. He came to study fossil fish; and Cuvier, convinced of his abilities, passed to him his own notes shortly before his death a few months after Agassiz's arrival in Paris.

Between 1833 and 1843 Agassiz brought out the five volumes of his *Recherches sur les poissons fossiles*, describing over 1,700 species and their relationships, which made his scientific reputation; and in 1837 he announced his conclusion that much of northern Europe had known an Ice Age in geologically very recent time, identifying boulders, valleys, and the positions and shapes of rocks as the effects of former glaciers. Much of what his predecessors had put down to the Deluge now received a better explanation; moreover, this seemed to confirm the beliefs of those who, like Cuvier, saw a succession of catastrophes in the Earth's history, and discontinuities in creation. In 1846 he left the University of Neuchâtel where he had been since 1832, and went to the USA first to the Lowell Institute in Boston and then, in 1847, to Harvard, where he entered on a public career in science, organising the collection of specimens and delighting and edifying the public with lectures and publications on natural history. He proved an inspiring figure to a whole generation; but the naturalists who had been so delighted to have so eminent a man of science settle among them, and could not but admire much of his work, found themselves less happy when he carried complicated theoretical discussions outside the ranks of the profession and before the general public. His old-fashioned and idealistic world-view, full of natural theology, left him isolated in the ten years before his death in 1873, though his popular following and prestige were undiminished.

In the *Essay on Classification*, Agassiz sought to bring before the world his vision of a universe resonant with pre-established harmony, the work of an intelligent and intelligible creator. He did not believe in any Great Chain of Being, arguing that Cuvier and his successors had clearly established the real existence of classes separate from one another that cannot be placed in any single line. He believed that the different systems of various authors were successive approximations to the real system in God's mind, indicating 'the identity of the operations of the human and the Divine intellect; especially when it is remembered to what an extraordinary degree many *a priori* conceptions relating to nature have in the end proved to agree with the reality, in spite of every objection at first offered by empiric observers.' For Agassiz, not only species, but also genera and the higher classes, had real existence 'manifested in material reality in a succession of individuals' which were comparatively very short lived. However long a species or a genus may have lasted from the creation of its first to the extinction of its last members, 'at all times these different divisions have stood in the same relation to one another and to their respective branches, and have always been represented upon our globe in the same manner, by a succession of ever renewed and short-lived individuals.'

The classes were therefore like the 'forms' of the Platonic philosophy, more real than the individuals that exemplified them. Cuvier's work on mummified cats from Egypt had established that 'there was not the shadow of a difference' between them and modern cats. The relation between successive faunas was like that between the works of successive schools of art; the earlier pictures have not changed into the later ones any more than species, themselves works of art made by God, have been transformed. Agassiz was more troubled about the relations between contemporary faunas: why North American creatures resembled those of Europe far more, for example, than those of Australia resembled those of Africa or of South America, even when they all inhabited a similar climate, was a mystery. He saw separate centres of creation for each zoological region, refusing to believe in any kind of evolutionary development that might have given similar creatures in different areas a common ancestor. All animals and plants, he believed 'have occupied from the beginning those natural boundaries within which they stand to one another in such harmonious relations. Pines have originated in forests, heaths in heathers, grasses in prairies, bees in hives, herrings in schools, buffaloes in herds, men in nations.' The European and American mallard seemed indistinguishable but must 'have originated simultaneously and separately in Europe and in America and ... all animals originated in vast numbers, indeed, in the average number characteristic of their species, over the whole of their geographical area, whether its surface be continuous or disconnected by sea, lakes, or rivers, or by differences of level above the sea, etc.'

Nobody could say that Agassiz failed to spell out the implications of his beliefs, and it comes as no surprise to find that Lyell said that Agassiz's essay, because of its extravagances, pushed him firmly into Darwinism. But in fact the difficulties of those who urged the diffusion of species from single centres where they had begun as a single pair, or those who urged descent with modification, were also formidable, and hypothetical bridges between continents (supposed to have later sunk into the sea) were proposed in the most off-handed manner. What Agassiz did succeed in doing was to polarise the views of those who believed in the world as a work of design, and those who saw it as governed by laws not very different from those of chance. His work represented the last flowering of the certainties of the epoch of Cuvier, before the coming of the dynamical world of Darwin where nothing was really fixed and any apparently stable state was no more than a temporary equilibrium – another teaching of *Natur philosophie*, but one that Agassiz must have abandoned when he sat at the feet of Cuvier.

Agassiz did not believe that the criterion of interbreeding would do for defining a species, believing that many domestic animals and cultivated plants have arisen from the crossing of distinct species; but that they 'are the result of the fostering care of man; they are the product of the limited influence and control the human mind has over organised beings, and not the free product of mere physical agents.' His cryptic remark about man having been created in nations would allow his readers to suppose that the inhabitants of the various continents might really be autochthonous or aboriginal, created in whole populations where they are now found; and while, like Mallards, the different races perhaps belong to the same species, all men are not brothers or cousins, springing from the same stock. This puts Agassiz some way from fundamentalist Christianity; and indeed while he makes ingenious use of religious rhetoric, alternating between emphasis on the intelligibility of God's design and its inscrutability, it would be difficult to reconcile his views with orthodox religion. The Deist with his belief in a First Cause might indeed take comfort in the evidence for the separate creation and design of each organism, and the real existence of species, genera and so on as works of art conceived by God; but to the orthodox Christian, well aware of evil and death and the need to fight the good fight, the struggle for existence has rather more to commend it as a model.

It is perhaps curious that in the second volume of the *Contributions*, at the end of the splendid illustrations of turtles and their embryology, there appear two handsome coloured plates illustrating variation in species. To a Darwinian, such differences indicate adaptation to a slightly different environment, or indicate how creatures differ so that natural selection will continuously eliminate the less fit. To Agassiz, actual creatures conformed more or less to the ideal form, and could not, except with continual intervention by man, diverge at all far from it; and varieties were therefore of no great significance, except for recognising a specimen, and defining the limits of a species. Agassiz also noted in his *Essay* points upon which Darwin was to seize: such as the resemblance between fossils in a given area and creatures now living there; the greater resemblance between extinct forms and the embryos of living creatures than between extinct and present-day adult animals; and the increase in complexity in more recent epochs as one follows the geological record. For Agassiz, none of these points had any evolutionary significance, but simply indicated the long-standing localisation of types, the way in which the oldest representatives in any class may be envisaged as 'embryonic types' of their orders or families among the living, and the way the creative Mind has worked. Agassiz referred to Oken, and to MacLeay and Swainson, as bringing before zoologists the analogies between creatures as well as their affinities; but he saw these as too little known and obscure to be taken as a basis for any kind of classifying, and his own system lacked the kind of tight unity that analogies had brought to these earlier classifications.

As well as meeting Cuvier, in Paris Agassiz had met and had come to admire Alexander von Humboldt. Humboldt was one of the towering figures in nineteenth-century science, having made his name with an expedition to Central and South America lasting five years from 1799. He there became especially interested in the distribution of organisms, particularly plants; where natural historians had tended to concentrate upon new or 'non-descript' plants and animals, Humboldt was concerned with the whole aspect of a flora and fauna, the dull or familiar species as well as the exotic from the European point of view. He brought back great numbers of plants and animals, as well as data on the geological structure and economies of the countries he visited, and great quantities of magnetic, meteorological, and astronomical observations. The whole of science seemed open to Humboldt; and he slowly published his materials in thirty magnificent volumes, exhausting his own fortune in the process, in Paris between 1805 and 1830. His expedition covered a large territory, and in the course of it he climbed nearly to the summit of Chimborazo in the Andes, which was then believed to be the highest mountain in the world; and at the back of his atlas of plates for the *Nova Genera et Species Plantarum* he included a picture to illustrate zones of vegetation – he was an innovator in visual language, being one of the first to realise that one can show more than topography on maps. At the left is Chimborazo, rising from the equatorial zone, with palm trees at the bottom and snow at the top and other distinct zones of vegetation going round it in stripes. Beside it comes Mont Blanc in the Alps, as a mountain in the temperate regions, which looks like the upper part of Chimborazo; and then on the right, Sulitelma in Lapland, in the Arctic region, like the upper part only of Mont Blanc. Going vertically upward corresponded to going further from the Equator.

For Humboldt, then, as for Agassiz (for whom he fixed up the post at Neuchâtel), geographical distribution was of great interest, although neither of them gave it an evolutionary significance or made very much use of it in classification. Where it did sometimes matter was in deciding whether similar populations geographically separated belonged to the same species or not. In the USA in the early nineteenth century, most naturalists were field workers and were keen to distinguish American populations. Even where, as with mallards, they differed only very slightly if at all from the type found and described in Europe (and it is of course the legacy of Linnaeus that northern European forms are taken as the typical members of groups), they tended to give it a specific rank and name. In that, in modern terms, they were describing a population that did not as a rule meet or interbreed with the European form, they had good reason to distinguish it; but in nineteenth-century terms, this made them 'splitters', applying a fine taxonomy and ending up with a long list of species. By the middle of the century, this approach had become unfashionable; it was recognised that individuals belonging to a species like Agassiz's turtles differed among themselves, and 'lumping' rather than splitting became the rule in taxonomy – unless there was good reason, very similar forms were put into the same species. Genera similarly were reduced in number. It was one of the great problems facing those like Agassiz who believed in the real existence of species and higher taxa that great authorities differed so widely in what they took to be actual species. The change in taxonomic fashion meant that some American naturalists like Thomas Nuttall and Titian Peale found themselves being cold-shouldered by more up-to-date biologists, especially as the initiative passed from the colourful individualist in the field to the professional attached to some institution.

Geographical distribution was easier to study in the plant kingdom, for the zones of vegetation that Humboldt found on mountains are easier to map than the boundaries between mobile animal populations; though work has been done on Birds of Paradise in New Guinea where species do occupy very restricted areas. In the natural system, the botanists preceded the zoologists; the great land-mark being perhaps the publication by A.-L. de Jussieu, nephew of Bertrand de Jussieu, of his *Genera Plantarum* in 1789, the year which saw the collapse of the *ancien regime*, and also the publication of White's *Selborne*. Jussieu's position at the Jardin du roi, and after the Revolution at the Museum gave him access to great collections; his family had moreover been eminent in botany for several generations and had built up a splendid library and herbarium of their own. His system caught on as the earlier natural system of Adanson never had; Adanson's was based on giving no special weight to any characters, whereas Jussieu's depended on finding characters of primary importance and general value, and thus like Aristotle gave especial importance to the reproductive organs. Whereas Linnaeus' 'sexual system' was based simply upon counting, Jussieu's depended more upon function; and particularly at the lower levels, he relied not upon any one character alone, but on a judgement of overall affinity. What Jussieu did was to provide a system, or 'method' as it was more politely described, that was both natural and workable, and

117

thus made the artificial system of Linnaeus obsolete.

The natural method was more slowly accepted in England, and indeed in the English-speaking world, partly through pragmatism (when there is a system that works, why bother to go for something different?) and partly because of the conservatism of the Linnean Society. The first full-scale botanical work in English to plump for it was the *Natural Arrangement of British Plants* of S.F.Gray, which came out in two volumes in 1821. Gray described himself on his title page as 'Lecturer on Botany, the Materia Medica, and Pharmaceutic Chemistry'; he was the son of a translator of Linnaeus, but had never accepted the artificial system. He lived by medical journalism, by writing works on pharmacy, and by lecturing at a botanical garden and a private medical school. He wrote the introductory parts, and his son (whose part was not publicly acknowledged) wrote the rest of the work. This son, John Edward Gray, became a most eminent zoologist, and eventually Keeper of Zoology at the British Museum; but at the outset of his career he had to endure the mortification of being blackballed at the Linnean Society in 1822 because of alleged disrespect for Smith, the president. As a result, Gray turned his attention from systematic botany to zoology, when at the Museum he wrote numerous works describing and classifying the various groups of animals in the collections.

Smith, in his *Grammar of Botany* of 1821 compared the systems of Linnaeus and of Jussieu, placing on his title page a quotation from Linnaeus: 'Natural Orders instruct us in the nature of plants; artificial ones teach us to kow one plant from another.' He wrote that Jussieu's was simply an essay towards a system, on which much more work needed to be done before such a mode of classification could 'serve the purposes of analytical investigation, to make out an unknown plant.' Smith does not mention Gray, but he does respectfully refer to Robert Brown, who was becoming the greatest botanist of the day, having circumnavigated Australia with Matthew Flinders on his famous survey of the coasts. He returned to write a most important essay on the Australian flora, and to become Librarian to Sir Joseph Banks. Brown like Humboldt did not only observe, but generalised; he was concerned with the distribution of plants, and with their classification, and was becoming convinced that an artificial system would no longer do. On the death of Banks, in 1820, Brown inherited for his lifetime his magnificent collections and library; but in 1827 he passed them over to the British Museum, becoming Keeper of Botany there.

The natural system was effectively promulgated by John Lindley in 1830 in his *Introduction to the Natural System of Botany*; he was Professor of Botany in the recently founded 'University of London' which was soon afterwards recognised and chartered as University College, and was a great promoter of botany and horticulture, and a prolific author. He felt that while there was undoubtedly still instability in botany, the natural system had so far prevailed that an introductory work was needed. The natural system was, on his view, based upon the principle 'that the affinities of plants may be determined by a consideration of all the points of resemblance between the various parts, properties, and qualities; and that thence an arrangement may be deduced in which those species will be placed next each other which have the greatest degree of relationship; and that consequently the quality or structure of an imperfectly known plant may be determined by those of another which is well known.' In contrast, in artificial systems the various facts were grouped without distinct relation to one another. In contrast to Smith, therefore, Lindley believed that a natural system would be, and in fact was, superior for practical purposes to an artificial one. Unlike many advocates of natural systems, Lindley did not believe that there were really in nature any groupings higher than species: a 'genus, order, or class, is therefore called natural, not because it exists in Nature, but because it comprehends species naturally resembling each other more than they resemble anything else.' In actually classifying, Lindley stressed the importance of physiological characters rather than mere modifications of external form.

One of the difficulties about the introduction of natural systems was that the naming of plants and animals which seemed to have been stabilised under the Linnean system once again became fluid. There was no longer agreement about the status of the higher taxa, which for Lindley were mere names, but for Agassiz real entities; and on numerous occasions the same organism was given a number of different names, either by those who knew of former ones but thought them inappropriate or by those who thought they were naming a creature for the first time. In nineteenth-century works, long lists of synonyms precede the description of any animal or plant, and as in the Renaissance erudition was as necessary for the taxonomist as observation. This was especially true of curious and exotic plants and animals, which had been described probably more than once from a poorly-prepared and preserved single specimen by somebody who had never seen it alive, and had no idea which characters were individual and which specific or generic. Some of the orchids of South America illustrated by Mutis and his team of artists had thus, by the time his plates finally came to be published, acquired many fresh names.

As large collections in capital or university cities became more the rule, and as careful consultation in libraries associated with them came to be expected of anybody daring to name an organism (and during the nineteenth century this privilege increasingly passed from the 'collector' to the classifier at home), so this kind of multiple naming became less common. Rules were drawn up giving the earliest name bestowed since the canonical editions of Linnaeus the priority; so that unless a name has already been given to something else or is otherwise invalid, the first name given is the appropriate one and all others are mere synonyms. These rules were first applied by national bodies like the British Association, but were then imposed universally by international agreement as international conferences to settle such questions as names and units became a feature of all the sciences after the middle of the nineteenth century.

A more fundamental reason for changing a name is that the old genus has been revised and the animal or plant is no longer judged to belong to it – perhaps the genus is indeed quite revised out of existence, being found not to bring together organisms now supposed to show close affinities. The generic name therefore has to be changed; but here the convention is that the first specific or trivial name stands, with the new generic name now forming the first part of the new binomial. Since splitters and lumpers are still with us, generic names are not fixed, and when one taxonomist has split up a genus a successor may reinstate it. Species are also sometimes split up, so that creatures supposed to belong to a single species are judged to belong to two or more very similar species. This again involves matters of judgement; when the species is split, then part of the old population has to be given a new name. The taxonomist has to make out a case for this, and his contemporaries and successors may feel that his new species is really just a race; in zoology trinomial names, the third word indicating the race, are used, but there are some who urge that below the species level there is a continuum of variation, and that it is a mistake to give names and thus the appearance of stability to what is really a state of flux.

The term 'subspecies' is found in the *Philosophy of Plants* of A.P. De Candolle and K. Sprengel, which was published in 1821 in its English version, and which represents a major support for the natural system, which was urged in the taxonomic parts of the book by De Candolle, and reinforced by Sprengel in the chemical, physiological and pathological sections – some of which make slightly odd reading, the book representing an amalgam of French rationalist classification and German *Naturphilosophie*. The book did not in this English translation make any great impression at the time, but points made in it were to become of significance in later discussions of the places of various organisms. Little was made of subspecies; but distribution was urged as 'fitted to afford the most important assistance in the *Classification* of plants', and the plant descriptions and diagnoses do contain much more careful and precise accounts of geographical limits than did most of those of the eighteenth century. There is considerable discussion of homologies, but no evolutionary theory.

While the botanists were perhaps most concerned with the introduction of the natural system, because the artificial system had been so useful in their field, and had to argue most forcefully for it; it was in zoology, and especially those branches concerned with invertebrates, that practical problems of classifying which involved theoretical decisions were particularly prominent. In France, Lamarck had worked on these groups, and so had Latreille whose book on the natural grouping of crustaceans, spiders, and insects came out in 1810. Latreille rejected any evolutionary speculation, and his book has admiring references to Cuvier, to whom as 'the restorer of zoology' it is dedicated. Latreille was insistent that any one character is insufficient for classification; and especially that external characters will not do. Just as Cuvier had used comparative anatomy in sorting out the mammals, so must the student of invertebrates use internal structures, though he must take into account metamorphoses, instincts and habits (that is, the whole life-cycle) also. His aim must be a simple and accurate ordering, displaying the majesty of nature.

Physiological and anatomical evidence (based on the organs of respiration) enabled Latreille to separate the crustaceans from the insects; and also, following Lamarck, to distinguish the arachnidae, which include spiders, ticks, centipedes and other not-much-loved creatures. While he agreed with Lamarck's separation of this group from the true insects, he disagreed with him about the characteristics of the two classes: in science as elsewhere it is possible to reach the same conclusion for different reasons. Arachnids had been put with the 'aptera' or wingless insects by Linnaeus; and Lamarck had followed his predecessors in making it a mark of the group that they did not, like the true insects, undergo metamorphosis. For Latreille the differences are structural; and he was able to point out some arachnids that do undergo metamorphoses (a centipede that begins with three pairs of legs, and successively acquires more until it has twelve, for example), and insects like the grasshoppers that do not. Latreille was convinced that nature, unlike grasshoppers, made no jumps, so that there were arachnids very like some groups of insects, and crustaceans very like some archnids, forming links in an unbroken sequence.

In the late eighteenth century and the early nineteenth, in France and also in Sweden where De Geer and Fabricius were working, much effort was being devoted to entomology; but in Britain the science was little carried on, and evoked the slightly comic figure of the man with a butterfly net. Butterflies were indeed admired and collected; and by the late eighteenth century the problem that had perplexed earlier entomologists like Mme Merian, that from similar pupae sometimes butterflies hatched and sometimes little wasp-like flies, was resolved when the flies were understood to be parasites – though it then became a problem for natural theologians. Some of the less admirable butterflies, like the Cabbage White, were of interest as agricultural pests; and William Curtis the botanist wrote in 1782 a monograph on the Brown-tail Moth which had gone in for a population explosion and stripped the forest and fruit trees of the London area of their leaves. Just as hunters have taught us much about foxes, tigers and grouse, so much has been found out about the life cycles and habits of insects by those trying to get rid of them; but work of this practical kind does not lead to natural classification.

Entomologists as a group tried on various occasions to form a scientific society and arouse enthusiasm in England in the early nineteenth century, a notable attempt being made in 1822 by a number of Fellows of the Linnean Society including MacLeay, and with J.E.Gray also there although he was not a Fellow. The final decision was to form a zoological club within the Linnean Society, so that Gray was excluded until an independent Entomological Club was formed in 1826. The first general treatise on entomology that aroused wide interest in the science was the *Introduction* by William Kirby and William Spence, which was published in four volumes between 1815 and 1826. This began as an informally written work in the form of letters, but the later volumes became drier and more systematic; altogether it is one of the classics of the science, and ensured that it was taken seriously. They urged entomology as a source of both utility and intellectual pleasure, and argued for the value of a natural classification for training the mind.

In their text, they pointed out some of the hazards which had made entomology interesting for their predecessors. Linnaeus for example had described the male and female of a kind of bee as belonging to different species because their antennae differed so much – the antennae of male insects often being, as they note, more complex than those of the females. In explanation they remark 'For what end the Creator has so distinguished them is not quite clear; but most probably this complex structure is for the purpose of receiving from the atmosphere information of the station of the female.' This is not put in quite the language of modern biology, but it indicates how a concern with classification can accompany, and indeed lead to, an interest in behaviour.

Kirby and Spence were impressed by MacLeay's work, but not fully converted; believing that he had been of great importance in drawing attention to the analogies as well as the affinities in the animal kingdom, and to the resemblances between the young of some species and the mature forms of others; but of Lamarck's theory of evolution that might be supposed to give some explanation of homologies and analogies they had nothing good to say, quoting one or two examples and adding 'It is grievous that this eminent zoologist, who in other respects stands at the head of his science, should patronize notions so confessedly absurd and childish.' When they came to differ from Cuvier, they expressed much greater diffience; but they did catch him out in certain mistakes where some characters did not in fact define a group. They made use of anatomy in their classification – or rather, we should speak of Kirby's, for the later volumes were written by him – and also chemical tests, in the form of heat and treatment with acids, to establish that the integument of insects differed from horn, shell or skin; and microscopic observations, for the rapid improvement in instruments at this time was bringing into view structures hitherto unsuspected.

Kirby and Spence were judicious rather than original in their classification, and were pragmatic in their wariness of patterns like those of MacLeay – believing that there was something in the idea of circles, but that five was only one of the numbers prominent in nature – though not going as far in this as Lindley with his refusal to see any group above the species as natural. Much the same line was followed by J.F.Stephens who in 1829 published, by subscription, a *Systematic Catalogue of British Insects*; the subscribers including MacLeay, Swainson, J.G.Children (who had got the British Museum post Swainson wanted), and the young Charles Darwin, then at Cambridge. Stephens worked at the Admiralty, but was given leave of absence in 1818 to work on the arrangement of the insect collection at the British Museum. He had therefore acquired familiarity with a large collection, and had also built up one of his own; but as an amateur he could remind Latreille that 'naturalists generally, however zealous in the pursuit of knowledge, have other and more paramount avocations to follow than those of attending to the minutiae of science, and that all are not equally fortunate with himself in being able to devote their exclusive attention to that branch thereof which their inclination prompts them to study,' and so their descriptions must be slower to appear and less full than he or they themselves might wish.

Stephens made the interesting point from MacLeay about the perennial dispute between splitters and lumpers, that writers were often guided by the extent of their collections, those with small ones not seeing the 'necessity for subdivision which those with larger ones think expedient'. This cannot be the whole story, for in botany later in the century, those like Joseph Hooker and Asa Gray, with access to magnificent collections at Kew and at Harvard, were lumpers; but there may often be something in it. Stephens made out a case for beginning with specific details and distinctions and then generalising rather than going the other way about, this being a typical English reaction to the spirit of system seen in foreigners – and also in MacLeay. His discussion of classification and his useful bibliography only occupy thirty pages of the book, the remaining 800 pages of which are taken with systematic lists of insects, forbidding reading indeed for any but the expert, and an indication that there were in England such people by this time.

Ten years after Stephens, John Obadiah Westwood wrote a more discursive work on *The Modern Classification of Insects*. He was an Anglo-Saxon scholar, and was a superb illustrator, especially of butterflies and moths. He drew plates that showed the life cycles of a number of close species and the plants on which they fed, and were thus both systematic and ecological. His classification book has small and neat illustrations done on woodcuts so as to fill the block and waste no space; where there is not room, he shows only one wing of a specimen, or a composite picture where the left side is the male and the right the female, or dotted lines to show magnified parts or how parts move. He was unfortunate in that he believed that J.V.Thompson's work on barnacles and other crustaceans had been 'demolished', and that this group had no radical metamorphoses, which for him was one of the great characteristics and problems of the insects. The book shows enormous erudition, which was duly rewarded when he became the first professor of entomology at Oxford in 1861; but it was not only the fruit of closet study, as he wrote:

> I have studied nature in the woods and fields, tending and observing insects in all their various transformations, well knowing, that the man who confines his researches to the mere collection and examination of museum specimens, can neither possess so intellectual an enjoyment, nor acquire so perfect a knowledge of the subject, as is to be derived from the examination of living nature; and it is both with pleasure and with pride that I now submit the results of my numerous observations to the reader.

Field work was both manly and gentlemanly, those great Victorian ideals, in a way that swotting in a library was not; but actual observation could assist classification, for although the convention is to classify adults, unless one had seen creatures through their metamorphoses one could not be certain about their life-cycles and their relationships.

Like his contemporaries, Westwood was keen to find connecting links or 'passages' where two groups meet, expecting nature to make no jumps. He noticed how the organs of insects are modified to suit their life, the length of tongue depending on the form of the flowers from which they collect honey for example; this was a further reason why a natural classification could not be based on a single character which might have been differently modified in two creatures otherwise showing many affinities. He insisted upon 'the rules of proportionate development', which seem to be a version of Cuvier's principle of correlation. He wrote of MacLeay's system with qualified approval, and in discussing the arrangement of the lepidoptera prints a diagram of circles, but this time grouped in sixes around a single one; this group was particularly confused at this time, with different authors in disagreement, reminding us once again that while all may agree on the desirability of a natural system, they may not agree at all about the details of an actual system which involves judgements and theories.

In 1835 Edward Newman had published a *Grammar of Entomology*, intended as an elementary work, which included a section on classification. In 1841 a new version of it came out; it is attractively illustrated and written. On taxonomy, he wrote:

> It may be said that the author should have given the views and arrangements of others in preference to his own. He would ask, whose system was he to select? That his own is the most simple and the most readily understood, no one will deny: that it is more perfect, or more accurate, or more philosophical, than any other, he does not presume to contend.

Newman's system was circular, based upon the number seven, one of the groups being central and the others grouped around it just as the cells in a honeycomb are arranged; this system could be reconciled with the approaches of Swammerdam, Linnaeus, Fabricius, and the eclectic Latreille with his multiple criteria. Newman was a noted naturalist of the day, and had first proposed his scheme in 1832; he was an all-rounder, publishing works on many branches of natural history and ending up as natural history editor of *The Field*. His little *Grammar* is of interest in showing the demand for an elementary work, and yet also the unsettled state of insect classification, where the most fundamental questions were still open.

Entomology had its particular problems, because of the vast number of kinds of insects, their small size, and the difficulty of being sure that what emerged from the pupa was really the insect one had followed from the egg and not a parasite or the parasite of a parasite. Among larger creatures the matter seemed more straightforward, and in ornithology, for example, while there were debates about splitting or lumping, and about the proposing of genera containing only one species, there was more general agreement about what went with what. The quinary system had been devised to cope with the problems of the insects and other invertebrates, and extended into birds by Swainson; but it never caught on widely, and hierarchical rather than circular arrangements were the norm in ornithology.

Latreille had used the anatomy of insects as the most weighty characters in his classification; and this was something that Willughby and Ray had begun with birds in the seventeenth century. This helped sometimes to sort out analogies from homologies, as when the swift was separated from the swallow; but further indications were desirable, and the ingenuity of taxonomists was shown in their use of new kinds of evidence.

Thus C.L.Nitzsch of Halle linked ornithology and entomology in the first quarter of the nineteenth century in his studies of the parasites of birds as an aid to classification. It had been recognised that certain parasites confined themselves to one species, and the phenomenon was made much of sometimes in discussions of providence, special creation, and evolution. What Nitzsch did was to study and describe the parasites of birds with the view that identical or similar parasites indicate close affinity; and this is indeed a line still followed as giving evidence as to a creature's place, though it seems a curiously roundabout way of proceeding, using the classification of parasites to determine that of their hosts. The principle can also be used in the insect world. Nitzsch also made careful studies of the distribution of the feathers of birds, a study called pterylography, finding that in the different groups the feathers are very differently arranged, and showing careful illustrations of various plucked birds. These pieces of evidence were used by him in addition to data from comparative anatomy, but he died before actually incorporating them into a system.

Among admirers of Nitzsch's work, including that on the carotid artery and the muscles of the larynx, was the austere Swede, C.J.Sundevall, who worked at the Stockholm museum, and developed his own system between 1835 and his death in 1875. Nevertheless, like Swainson he valued the external characters of birds at least as high as the internal, and indeed in practice rather higher, so that his system in its foundations ran counter to the whole tradition of nineteenth-century biology. He coined new names for the higher taxa as well; and since his system separated the chats from the thrushes, and united other groups usually widely separated, it aroused considerable interest but won few converts. The dominance of internal parts was generally accepted as the basis for classifying, at any rate among the experts, throughout the nineteenth century, and since.

At a more popular level the more profound classification of organisms may not be as successful as a more intelligible if more superficial work; and this was well illustrated in 1837, when two works on *British Birds* began to appear; one by William Yarrell proved very successful and in later editions became the standard work for the nineteenth century, although (perhaps even because) it had no claims to great originality but was judicious and agreeably illustrated with woodcuts. The other was eventually about twice as bulky, took longer to come out, but was more important for its theory of classification; it was by William Macgillivray, of Edinburgh, who had helped Audubon with the technical parts of his *Ornithological Biography*. It too had illustrations, but the most splendid are dissections, and in general just the heads of the different birds are shown, which is much less useful to the ordinary birdwatcher. Macgillivray saw the difference in the organs of voice that separate the passerine birds of the Old World from those of the New, but his chief interest was in the digestive organs. He wrote:

> The physiology of these organs forms a most convenient centre of relations, affording, as it were, a key to the more intelligible functions, and determining the food, the haunts, the flight, the mode of walking, and other actions of the bird. It also throws much light upon the affinities of groups, and tends to prevent the frequently absurd associations imagined by persons who form systems by arranging birds' skins on their parlour floors.

These remarks were addressed to those who would rely upon external characters and confine their work to the closet.

Macgillivray's five volumes came out over fifteen years, and were never as popular as they perhaps deserved to be. His emphasis upon the digestive system was qualified in practice, but one can say that along with Nitzsch he introduced new characters important in classification. While not arranging the birds in an hierarchical order, he considered the crow family as probably the highest, with evidence from their adaptability and intelligence and also from their anatomy and physiology; and in this conclusion he has been followed by many ornithologists who have thus been prepared to dethrone the eagle in favour of a rather more democratic philosopher-king of the birds. All in all, Macgillivray's book shows an acute and independent man applying internal characters as well as external in classifying birds in a discursive and readable way, and it deserves to be better known.

Macgillivray's last volume came out in 1852, and thus brings us close to the epoch of the *Origin of Species* which was published in 1859. During the seventy years that separate Jussieu's book from Darwin's, the natural method had become universally accepted; but whereas Linnaeus' artificial system was a scheme to be applied as it stood, like a convention on naming, the natural system was an ideal to which any actual arrangement was a more or less good approximation. There was no one natural system; all that one was committed to was the belief that there was some real order in nature, and that this would probably be found out when a range of characters were properly weighted and taken into account. Some like Agassiz saw the order as an expression of God's mind, with every division imposed at the creation; others like Lindley or Macgillivray saw species as real and everything else as artificial; others worried about the status of species and sub-species. Some schemes had a numerical basis, but outside mineralogy where the laws of crystallography gave a mathematical basis, this proved of little ultimate value in natural history; but ever since Euclid, our culture has been disposed to assign an excessively high value to mathematics, and numerical speculations remained seductive.

It should not be supposed that the urge to classify was confined to those working in natural history, although that has been the topic of recent chapters; and before going on to look at how Darwin's theory modified theories of classification of organisms, we shall turn to classifications in other fields. The eighteenth and nineteenth centuries were great periods for schemes of classification, and the urge to impose order was strong in fields closer to the concerns of everybody than are ornithology or entomology. No doubt, too, the successes of Linnaeus, of Buffon, of Cuvier, and others encouraged those who believed that the main task of science, meaning either the study of the world or the 'moral sciences' which were concerned with man, was to classify – this being not a tedious business done with dessicated corpses on the parlour floor, but an exciting way of finding out how everything was really arranged and fitted together. Hegel wrote that when science paints its grey on grey, a form of life is over; the owl of Minerva flies only with the coming of the dark. But his contemporaries for the most part would not have agreed; their concern was with the living and not the dead, whether birds or ideas were their subject.

6

Everything in its place

To the taxonomist, the problem is to determine the natural groups into which things fall, and then to devise some simple key so that anybody else can, without spending the same time in profound study, put them into the right category. When the things are animals or plants in their astonishing variety and with their complex affinities and analogies, then we have the science of natural history; but similar problems meet us when we confront the inorganic world, and also when we try to deal with human artifacts, whether material, like arrowheads or nuts and bolts, or intellectual, like sciences, or midway between, like books. The development of indexes, inventories, catalogues, and systems of knowledge is an indication that cataloguing and ordering really is a fundamental preoccupation and not something confined to those who take an unusual interest in the natural world. Anybody who has ever used an encyclopedia, or a library catalogue, or the membership list of a society, or has hunted frustratingly round a department store, will be aware that different classifications suit different purposes, and that some systems are a good deal better than others.

The natural historians of the Renaissance, such as Aldrovandi and Gesner, are sometimes described as encyclopedic naturalists or even simply as encyclopedists. Like Pliny in ancient Rome, they sought to cover the whole range of knowledge within their subject, and this brought them up against the problem of organising their bulky volumes. In our century, too many people do not bother to read books right through, dipping into books written to be read, as though they were reference books to be consulted; but encyclopedias really are reference books, and the reader needs to be able to find his way rapidly to the section he wants. In general, encyclopedias ever since the Renaissance have followed the system of putting entries in alphabetical order. In Topsell's compilation on *Fore-footed Beastes*, for example, this is what is done, each volume (the two later ones in the 1658 edition being on 'serpents' and 'insects') covering its field from A to Z. A difficulty about this is that an animal may have various names where it could be found – bull, cow, kine, ox, for example – and that the entries for the dog and the wolf which one might have wanted to compare are far apart. The book has no index of names of creatures, but it does have an 'index of chapters' (a table of contents) at the back, and also a 'physical index', which 'contains plentiful Remedies for all Diseases incident to the Body of Man' prepared from the various creatures described in the text. The preacher would have to dip about in search of edifying anecdotes, but the apothecary or physician could look up 'Dragons bitings cured' or 'Guts wringing' and find what to do.

When books were organised rather more systematically, like John Parkinson's *Paradisus Terrestris* of 1629 which describes in English garden flowers, plants for the kitchen garden, and trees for the orchard, better indexing was required, and his book has three: one of Latin names, another of English ones, and a third of 'Vertues and Properties'. Many works of the seventeenth century (and later) though loosely organised and hard to find one's way around, had no index, but usually had a table of contents; an example is Robert Hooke's *Micrographia* (1665) which is celebrated for its enlarged plate of a flea seen through a microscope but contains descritions of many experiments and observations in several sciences. Only in eighteenth-century reprints, which were abridged anyway, was an index provided, and this may well be why, although the book was famous, some of the ideas in it (on the composition of crystals from spherical atoms, for example) were unnoticed until they were later proposed by others. The voluminous and unordered writings of Hooke's patron, Robert Boyle, fared better because in the eighteenth century they were collected and published as a set of five volumes (later reprinted in six) with an index.

Writings like those of Boyle and Hooke are not unlike a scientific journal in that they described numerous distinct pieces of work, though naturally they have rather more of a theme than do issues of a journal as a rule. The scientific journal transformed science from the 1660s because it meant that new discoveries could be made known to the scientific community speedily, and before there were enough of them to fill a book. The early volumes of the Royal Society's *Philosophical Transactions* were not indexed, but in the eighteenth century various cumulative indexes were made and the volumes then coming out also acquired indexes. Without such helps, items published in journals become very hard to dig out after a few years. It is not always easy to find the index even if it is there; thus in Topsell's volumes, as in many French works of the nineteenth century, the table of contents is at the back, while in some journals and in various nineteenth-century English books the index is to be found at the front. The editors of a recent reprint of William Whewell's *History of the Inductive Sciences* (1837) seem to have missed the index at the beginning of volume one, and have painstakingly prepared another very similar one at the back of volume three.

Where a book has had a long history of importance, it may well be that the original index prepared by the author or some contemporary no longer meets the needs of most readers, so that a modern index is more valuable – one might even like to have both. Anybody who reads a fair number of books old or new will know that indexes never quite meet one's needs, and there is really nothing for it but to make one's own on the endpapers or the half-title – and that after a few years this too will seem obtuse. An author or his indexer may have supposed that he was describing an election in a society or an experiment; to the twentieth-century reader he may be saying something about professional status or instrumental accuracy, but it will not be indexed under such headings. A good reason for owning books rather than borrowing them is that one can index and annotate them oneself, for librarians only prize old annotations in their copies.

An index represents an alternative ordering of the material in the book, while the table of contents is a summary of the actual ordering. This is clearest when the index is a thematic one: Topsell or Parkinson could have written their volumes as medical works, taking diseases in some kind of order and listing remedies. Their index then might have been of the names of animals or plants. This approach might have been impracticable, and would have resulted in a very different sort of book, but it represents an ordering that was possible in principle. Even indexes of people's names represent an alternative ordering, for the matter in the book might have been grouped under names in a kind of biographical dictionary, while an index of Latin names indicates that the book could have been written in that language when things would perhaps have been described in a different order. Because there is no limit to the number of ways in which we might order material, there is no ideal index; which does not stop us from saying, as of classifications of plants or animals, that some are not better than others.

The diseases in Topsell and Parkinson, and in more-official publications like the Bills of Mortality published in seventeenth-century London, make rather odd reading to us. Physicians in the seventeenth, as in the twentieth, century cannot have much needed specifics for curing the bites of dragons; but they must have seen those with diseases to which they gave names, like rising of the lights or mother-fits, which are not found in our medical statistics. The classification of diseases is something which has been a major part of medicine throughout most of its history; and associated with, or sometimes opposed to, it has been the classification of patients. Here much more obviously even than in natural history, reclassification has been associated with changes, often fundamental ones, in theories.

Diseases can be seen as an imbalance within the patient, or as an invasion by some agent, material or immaterial, or as a bit of both. In Greek medicine, the emphasis was on the balance of humours (blood, phlegm, and yellow and black bile) within each person, and illness was thought of as an excess or defect of one of them. They were associated with the four elements, earth, air, fire and water; and people fell into four main groups or temperaments, being sanguine, phlegmatic, bilious or melancholy according to the particular equilibrium of humours appropriate to them. Persons of different temperaments were liable to different sorts of illness, and the balance of their elements had to be restored in different ways. There was not much point in classifying different kinds of imbalance with great care; the important thing was to recognise to which group a patient belonged.

We all know that mild diseases do take different people differently, and that one person in a family escapes something that has laid all the rest low; but acute diseases do seem to take much the same course with everybody, and there is a good description in Hippocrates of an epidemic of mumps. This is clearly a disease that can be classified because it affects all patients in much the same way, and is therefore easier to interpret as an invasion than as an imbalance. But such interpretations seemed bound to be hypothetical, and illnesses were often simply described as a syndrome, a bundle of symptoms that went together and for which some treatment or other could be recommended. The disease is therefore given a name, but the account of it is purely empirical and inductive; it is simply distinguished from other species of disease for which different treatment is appropriate, and some one symptom may be recognised as a key for diagnosis.

Modern medicine still has its syndromes; but, just as in natural history species were formerly defined in terms of external characters, and latterly and more certainly and naturally in terms of internal ones, so diseases have ceased to be defined by external symptoms and are now characterised when possible in terms of the internal organs affected and the agent deemed responsible for causing the disease. The great change here may be said to have begun with the growth of hospitals associated with medical training towards the end of the eighteenth century, so that doctors saw far more examples of diseases than they would have in ordinary practice and could learn pathology through autopsies; then, in the mid-nineteenth century, the germ theory of disease was proposed by Pasteur, Koch and others. The classification of patients ceased to seem so interesting, and they became mere examples of scurvy, cholera or tuberculosis, despite Coleridge's coining of the useful term 'psychosomatic' in the early nineteenth century. In our century, the conquest of some of the most formidable diseases, and the rise of genetics, has perhaps begun to shift attention back to the patient's individuality, though a large hospital is not on the whole a good place to investigate or display it.

One of the problems about the classification of diseases is that, over time, they seem to change their effects. The mumps described in the Hippocratic writings is unmistakeable, and that disease must (like the Egyptian cats dissected by Cuvier) have remained little changed through more than two thousand years. In the same writings, varieties of intermittant fevers are described and distinguished that to moderns would all simply seem to be malaria: to the Greeks, and no doubt to the patient, the frequency of the attacks is significant and a 'tertian' is different from a 'quartan' ague, but in modern medicine these are like colour differences within a species and not very interesting for taxonomy. While mumps has gone on unchanged, some diseases have not, and there is much debate about the nature of the plague at Athens carefully described by Thucydides the historian in the fifth century BC. This does not seem quite like any plague now afflicting mankind, and it may be just have been different. In the last hundred years scarletina has ceased to be a terrifying disease with a high death-rate; and in a shorter period influenza has come in cycles, slightly different each time, so that it seems to be a genus rather than a species of disease.

Not only do we have a theoretical change, then, so that malaria is no longer attributed to bad air or influenza to a mysterious malign influence from the Devil or the stars, but it also seems as though we are engaged in classifying unstable entities. Diseases change not only over space, so that different airs, waters and places have different illnesses associated with them as the Greeks knew, but also over time, so that the medical taxonomist is working on something in flux. This may seem alarming, but meteorologists classify clouds, and it is something that can be lived with; and indeed some explanation in terms of mutations and of natural selection can even be given for it – precisely the same problem confronted natural historians as the Darwinian theory became accepted.

Medical men were not alone either in classifying syndromes, where a number of symptoms were observable but their cause was unknown. Philosophers who follow Hume would argue that all science must be like that, which seems rather extravagent; but in chemistry in the nineteenth century, for example, the various elements were classified in groups for the purposes of qualitative analysis in very much this way. The analyses were done by, and are still taught to, those without understanding of the principles of chemical equilibria; but if one knew that solutions containing certain metals gave a precipitate with hydrogen sulphide in acid solution, and others in alkaline solution, and so on, it was possible to use such tests to form classes, to determine in any given case to which class the unknown metal in the substance to be analysed belonged, and then to do spot tests to identify which member of the class it was. The system could be systematically set out in tables, as was done for instance in *The Chemical Atlas* of A. Normandy in 1857 and in countless textbooks since.

These tables were based upon clusters of properties rather than fundamental understanding of chemical affinities, and their aim was diagnosis rather than a natural arrangement; they formed a useful key, and they were in use many years before a satisfactory table of chemical elements was devised. Lavoisier in the 1780s had defined a chemical element as a substance that could not be decomposed; certain of his elements were in after years decomposed, and other elements were discovered as chemical analysis was improved and new methods such as spectroscopy were invented, so that the number slowly climbed towards about ninety – a very small number compared to the number of species that natural historians had to deal with. In the second decade of the nineteenth century, A.M.Ampère tried to devise a natural system for the arrangement of the elements, based like those of his contemporaries at the Museum d'Histoire Naturelle on multiple characters and on arranging elements in families, but this never proved more than a curiosity.

At the same period, Davy who had recognised that sodium and potassium (which he had discovered) belonged to the same family, urged that chlorine and iodine were also closely related; and other groups of elements were distinguished during the first half of the nineteenth century on chemical grounds. Just as Davy was working on potassium, Dalton in Manchester was proposing his version of the atomic theory in which each element was composed of atoms irreducibly different from those of any other. These atoms were characterised by weight; and Dalton with his training in mathematics thus brought into chemistry the hope of some sort of quantification, or anyway of classification based on numbers.

Eight ounces, pounds or tons of oxygen combines with one of hydrogen to form nine of water; so if, as Dalton assumed for simplicity, water is composed of one atom of each element (HO), then the atom of oxygen will weigh eight times as much as the atom of hydrogen. On the other hand, two volumes of hydrogen combine with one of oxygen; and this inclined Davy, Ampère and Avogadro to assign water a formula like H_2O. If it contains two hydrogen atoms, then to keep the weight ratio right, each oxygen atom must weigh sixteen times as much as a hydrogen one. There was, for half a century, no way of knowing which authority was right; and for many other compounds there was just the same kind of uncertainty. From similar analyses and syntheses, different authors came up with different series of formulae and atomic weights.

It was not until 1860 that an international conference (one of the first of the kind) was called at Karlsruhe to settle the question; it failed in this task, but Cannizzaro's suggestions were read by participants on the way home and his convention, based on the ideas of his countryman Avogadro published about half a century before, was soon afterwards adopted, and led to the formulae and atomic weights that we now use. No previous series of atomic weights led to any general grouping of the chemical elements; within families some regularities were visible but that was all. The new figures enabled a number of people in different countries to arrange all the elements in a table in which both the numerical and the chemical characters were taken into account. The best known of these was Mendeleev, whose first 'periodic table' was published in 1869. He noticed that if one arranged the elements in order of increasing atomic weight, then related elements follow one another periodically. At first he put them in vertical columns, when the related elements fell in horizontal lines, and later he turned it the other way into the form now familiar as an adornment on the walls of chemistry lecture rooms. In such a two-dimensional system, the vertical, horizontal, and diagonal relationships are all significant.

			$Ti = 50$	$Zr = 90$	$? = 180$
			$V = 51$	$Nb = 94$	$Ta = 182$
			$Cr = 52$	$Mo = 96$	$W = 186$
			$Mn = 55$	$Rh = 104,4$	$Pt = 197,4$
			$Fe = 56$	$Ru = 104,4$	$Ir = 198$
		$Ni = Co = 59$		$Pd = 106,6$	$Os = 199$
$H = 1$			$Cu = 63,4$	$Ag = 108$	$Hg = 200$
	$Be = 9,4$	$Mg = 24$	$Zn = 65,2$	$Cd = 112$	
	$B = 11$	$Al = 27,4$	$? = 68$	$Ur = 116$	$Au = 197?$
	$C = 12$	$Si = 28$	$? = 70$	$Sn = 118$	
	$N = 14$	$P = 31$	$As = 75$	$Sb = 122$	
	$O = 16$	$S = 32$	$Se = 79,4$	$Te = 128?$	
	$F = 19$	$Cl = 35,5$	$Br = 80$	$J = 127$	
$Li = 7$	$Na = 23$	$K = 39$	$Rb = 85,4$	$Cs = 133$	$Tl = 204$
		$Ca = 40$	$Sr = 87,6$	$Ba = 137$	$Pb = 207$
		$? = 45$	$Ce = 92$		
		$?Er = 56$	$La = 94$		
		$?Yt = 60$	$Di = 95$		
		$?In = 75,6$	$Th = 118?$		

Mendeleev's first Periodic Table, in which the elements are arranged so that the families are horizontal; predicted ones are indicated by a question mark. The numbers are the atomic weights. From *Zeitschrift für Chemie*, 12(1869)405.

Mendeleev staked his reputation on his table as none of the others working on it did, and it is this which entitles him to be seen as the Linnaeus of chemistry; as Freud observed, it is one thing to become acquainted with an idea, and another to get married to it. Perhaps it is surprising that the greatest achievement of this physical science in the nineteenth century should have been a taxonomic one, bringing order into what had previously been a formidable collection of facts. When one of Mendeleev's rivals, Newlands, urged a somewhat similar table upon the Chemical Society of London a year or two earlier, a distinguished member of his audience asked him if he had tried arranging the elements in alphabetical order, evidently believing that any arrangement would be justifiable with a certain amount of ingenuity, and that a truly natural order remained to be found.

Newlands' table of 'octaves' was indeed very imperfect and he had not quite got the key to the problem of arranging the elements; but any system has to be justified, especially if sceptical contemporaries are to be convinced that it is more than artificial, like the analysis tables or the *Synoptic Tables of Chemistry* designed to be put up on a wall that A.F.Fourcroy published in 1800, which were to serve as a summary of his lectures in Paris and which again are reminiscent of Linnaeus' *Systema Naturae* of 1735. Similarly the idea that chemical compounds were exemplars of relatively-few kinds of ideal 'types' was a popular and artificial way of classifying them in the mid nineteenth century, and is sometimes misleadingly described as a theory rather than a taxonomic device. Mendeleev justified his system by using it to make predictions; for on putting the elements into their families he found that gaps were left. The properties of the elements around the gap could be used to predict those of the missing element, and some of Mendeleev's predictions turned out to be astonishingly accurate. The mixture of the qualitative and the numerical aspects in his system meant that the gaps were more obvious than those which natural historians had left in their systems for 'missing links' of various kinds, but the idea was not wholly new. When in doubt, as over the elements tellurium and iodine, as to the exact order of elements, Mendeleev preferred chemical data to atomic weight; so that while atomic weights were crucial to the system, one might feel that had chemists been more familiar with natural historians they might have had a workable system rather earlier.

Even in the highly mathematical science of astronomy, classification played an important role in the middle of the nineteenth century with attempts to understand stars and nebulae. Nebulae were classified by their general shape by William Herschel, and through the nineteenth century there were continual efforts to establish whether they were composed of gas or of many small or distant stars (as Galileo had shown the Milky Way to be); whether or not some of them were solar systems in the making; and whether they were a part of our galaxy or belonged to independent arrangements. The spectroscope answered some of these questions, indicating that some nebulae were gaseous; and it also helped astronomers to classify stars rather more accurately than in terms of colour (red, white or blue). Astronomers like Secchi in Italy and Lockyer in England put the stars into classes, and Lockyer even tried to give an evolutionary significance to the classes in the last decades of the century.

The classification of diseases, chemical elements, and stars is clearly a part of science understood in a narrow sense. But a favourite preoccupation of philosophers professional and amateur was the classification of sciences; and here one is no longer ordering nature, but classifying intellectual artefacts. Ampère tried his hand at such a classification, in an essay on the philosophy of science published in 1834 when he had already achieved a great reputation for his work on electricity. The sciences he divided into two kingdoms, the cosmological and the noological sciences (those of nature and of mind); and the kingdoms are divided into branches, sub-branches, and then sciences of the first, second, and third orders. The kingdoms roughly separate what we would call pure and applied science on the one hand, and arts and social science on the other; there is complete symmetry, with eighty-four sciences of the third order in each kingdom. The system is like Fleming's in zoology, a rigidly bifurcating one, so that 'chemistry' is one of the two species (with 'experimental physics') of 'elementary general physics', which is one of the two second order sciences that compose 'general physics'. This comes in the sub-branch 'physics in the strict sense', belonging to the branch' physical science' of the sub-kingdom 'cosmology in the strict sense' and the cosmological kingdom. The system is clear but seems more appropriate to a library arrangement than to a real division of knowledge, because the connections in the intellectual world are more complicated than a bifurcating system can show – thus pharmacy, molecular mechanics and mineralogy appear far from chemistry. It indicates, too, how such a system can soon go out of date, for chemistry is a science of the third rank while zoology is a sub-branch having eight sciences equivalent in importance to chemistry grouped under it.

Similar curious features are to be found on the 'arts' side of Ampère's classification, which never gained adherents in any numbers. But in Ampère's day there were others trying to arrive at a natural classification of knowledge in a different philosophical tradition going back to Kant at the end of the eighteenth century. Descartes in the first half of the seventeenth century had proposed a mechanical world view, in which anything was properly explained only when the mechanism that brought it about was disclosed. In a popular image, the world was compared to an enormous clock, and God to a super-clockmaker who had made a mechanism that never needed setting or winding up. This idea retained currency into the early nineteenth century, being made use of by Paley in his famous *Natural Theology* of 1802 which was made a compulsory book in some English universities for many years and where, in a powerful opening passage, the world is compared to a watch. If the world is a mechanism, then man too is a kind of clock; and one of Descartes' followers in the eighteenth century wrote a famous book on *L'Homme Machine*. For Descartes himself, man was distinguished from animals and other machines by having an immortal soul – the 'ghost in the machine' of hostile delineators – but to mechanists this distinction became of decreasing importance as time went on.

To the Cartesian, then, mechanics was the basic science, and all others should in time be 'reduced' to it, and explained in terms of pushes and pulls, and collisions of fundamental particles. There was, therefore, a hierarchy of sciences, with physics the most fundamental and other sciences in varying degrees approaching it as they were more mechanical and mathematical. The difficulty was that the most advanced and magnificent science, astronomy, did not fit this picture; for Newton's world was one in which the Sun and the Earth attracted each other across void space, and was therefore mechanically unintelligible although it was splendidly mathematical. To Kant and his followers in Germany, force rather than matter was the underlying reality, and the reduction of sciences was to be firmly opposed. The various sciences were giving explanations on different levels, and they depended upon different fundamental concepts so that zoology could not be reduced to physics without leaving out the really important parts of it.

These ideas fitted in with Romantic emphasis upon individuality, and with the new emphasis upon history where the analysis of the very different pasts of different countries and peoples was coming to replace the assumption that all nations would simply follow more or less rapidly in the path of the most civilised, probably France. There was no paradigm history, and no paradigm science; and the emphasis on force, which later came to be called energy and defined more precisely, also led to renewed interest in the life sciences because life was a manifestation of energy. Mechanics was merely the science of dead matter, of billiard balls and beams, and thus occupied the bottom rank of the hierarchy rather than the top; at the top came the sciences of man, then the other life sciences, then chemistry and those like electricity and magnetism concerned with forces.

Such a scheme was presented by Hegel in his *Encyclopedia* which appeared in various editions after 1817. Two translations into English of this work on the philosophy of nature have recently come out; and they show how well-informed Hegel was about the science of his day, and how interested he was in getting right the various levels at which different sciences explained phenomena. There was one notorious point at which Hegel seemed to desert his own principles, and that was in his account of colour. Newton and those working in optics after him had sought to account for colour in terms of the motions of particles or waves of light, following a classic programme of reduction. To Goethe, Hegel's older contemporary, this meant that all the interesting questions about how we see colours were neglected, and he launched into a formidable polemic against the Newtonians. In a scheme of hierarchies and levels, it would have been possible for Hegel to say that both parties were right; that the Newtonians were doing physical optics, while Goethe was doing psychological optics, and that the questions asked and the answers that were appropriate in the two sciences were different – but in fact Hegel set the two in competition, and came down on the side of his countryman.

Hegel's full scheme did not appear in print until 1842, long after his death and when the science had already become obsolete – for anyone classifying sciences is certainly classifying a flux – and had little contemporary influence outside the ranks of the Hegelians. In England a Kantian scheme for the ordering of the sciences was proposed by William Whewell from 1837 in his important books on the *History* and the *Philosophy of the Inductive Sciences*. Whewell was to become Master of Trinity College, Cambridge, and was one of the great polymaths of nineteenth-century England, of whom it was said that 'science was his forte, and omniscience his foible'. Becoming aware of German thought, he was concerned to expound a new kind of natural theology different from that of Paley; for he recognised, with Kant, that one cannot *prove* the existence and goodness of God from the facts of nature. For him, the right way was to assume God's existence, which thus made the world intelligible and science possible; and then to refine the original idea of God by comparing it with what one learned about Him through the creation. In the same way, each science had its basic idea which had to be accepted at the outset, and which was then refined by experiment, observation, and theory. Real science was the outcome of the interaction of active minds and nature, and not just a collection of facts observed without preconceptions. Since each science had its own idea, and could only make progress when the idea was found and stuck to, reduction was impossible; and Whewell was on this ground a firm critic of the introduction of atomic theories into chemistry, which was not a mechanical science but one based on the idea of analysis. Chemistry in Whewell's scheme occupied a place between the 'mechanico-chemical' sciences of electricity and magnetism, and the 'analytico-classificatory' science of mineralogy.

Whewell was one of the first to get German philosophical ideas taken seriously by men of science in Britain, who had previously espoused in public what they took to be a 'Baconian' way of proceeding, in which with open mind the observer or experimenter collected facts and generalised from them, not attaching any weight to theories and hypotheses but looking for laws. In this spirit the Chemical Society of London, founded just as Whewell was bringing out his books, refused to publish in its *Journal* papers of a purely theoretical character. But he was not the first to look to Germany; for at the very beginning of the century the poet and philosopher S.T.Coleridge, for some years a close friend of Davy, had followed Schelling in seeing chemistry as a science not of mechanisms and matter but of polar forces, and had rejoiced when

Ordering the World

CLASSIFICATION OF SCIENCES.   117

Fundamental Ideas or Conceptions.	Sciences.	Classification.
Space	Geometry	
Time		Pure Mathematical Sciences.
Number	Arithmetic	
Sign	Algebra	
Limit	Differentials	
Motion	Pure Mechanism	} Pure Motional Sciences
	Formal Astronomy	
Cause		
Force	Statics	
Matter	Dynamics	
Inertia	Hydrostatics	Mechanical Sciences.
Fluid Pressure	Hydrodynamics	
	Physical Astronomy	
Outness		
Medium *of Sensation*	Acoustics	
Intensity *of Qualities*	Formal Optics	Secondary Mechanical Sciences.
Scales of Qualities	Physical Optics	
	Thermotics	(Physics.)
	Atmology	
Polarity	Electricity	Analytico-Mechanical Sciences.
	Magnetism	
	Galvanism	(Physics.)
Element (*Composition*)		
Chemical Affinity		
Substance (*Atoms*)	Chemistry	Analytical Science.
Symmetry	Crystallography	} Analytico-Classificatory Sciences.
Likeness	Systematic Mineralogy	
Degrees of Likeness	Systematic Botany	
	Systematic Zoology	} Classificatory Sciences.
Natural Affinity	Comparative Anatomy	
(*Vital Powers*)		
Assimilation		
Irritability		
(*Organization*)	Biology	Organical Sciences.
Final Cause		
Instinct		
Emotion	Psychology	
Thought		
Historical Causation	Geology	
	Distribution of Plants and Animals	
	Glossology	Palætiological Sciences.
	Ethnography	
First Cause	Natural Theology.	

The arrangement of the sciences, from W. Whewell, *Philosophy of the Inductive Sciences*, 2nd ed., 1847, II, p.117.

Davy had shown that chemical affinity was electrical, and that chemical properties were modified by electric charges.

Coleridge was the moving spirit behind the plan to bring out an encyclopedia in English that would not be organised in alphabetical order like previous ones had been, but in a 'philosophical' arrangement like that at which Hegel had aimed. The outcome of this was the *Encyclopedia Metropolitana*, with an introduction by Coleridge, which came out in twenty-nine volumes between 1817 and 1845. The introduction was mangled by editors, and the original was separately published by Coleridge in his periodical, *The Friend*, in 1818. The idea was to have book-length treatises on the various branches of knowledge arranged in logical order, associated with a large-scale dictionary. What happened was that there were two big volumes on the pure sciences (containing no empirical component) of grammar, logic, mathematics, theology, law, and ethical and metaphysical philosophy – some of which seem strange bedfellows. Eminent men wrote the treatises in this and other sections, including F.D.Maurice the great theologian on ethics, Richard Whately on logic, and George Peacock and Augustus de Morgan on branches of mathematics.

The pure sciences are followed by six volumes on the 'mixed sciences', those with an empirical component, which include not only what we think of as sciences but also sculpture, music, poetry, agriculture, fortification, and commerce. In England at this time the word 'science' meant any organised body of knowledge, and therefore included arts subjects; any divide into two cultures, as in Ampère's classification, was still in the future. The difficulty was that although the various treatises were often very good indeed, and some of them were republished in book form and became famous, they do not hang together very successfully. The transitions and levels about which Whewell and Hegel tried to be careful were not delimited; and while in principle it would be possible, given a great deal of time, to read the *Metropolitana* volumes right through and get thereby a good general education, the reader might be perplexed to find that between 'painting' and 'engraving' he has to go through 'heraldry', 'numismatics', 'poetry', and 'music'. In ordering sciences as in ordering animals, any linear sequence is bound to display anomalies, but this one seems odder than necessary.

The mixed sciences were followed by five volumes on history and biography; history being seen as the doings of men rather than, in the Enlightenment manner, as the working out of inexorable laws like those of astronomy. These tomes were then followed by twelve 'miscellaneous and lexicographical' ones, which like an ordinary encyclopedia have short entries on numerous scientific words, which in this case are not cross-referenced to the treatises; one has to use the last volume, the index, which follows three of plates, to find one's way from one part to another. The general plan is thus not entirely dissimilar to the current *Encyclopedia Britannica*, with some volumes of short entries and others with longer articles; but the form of the *Metropolitana* proved too unwieldy, it took so long to come out that much was obsolete before all was published, and it was very difficult to think how to revise such a set. Its fate meant that the 'philosophical encyclopedia' organised in a natural rather than alphabetical order never caught on in the English-speaking world, where there was a long-standing suspicion of systems anyway.

Germany had its systematic ecyclopedias, but it was in France that the most famous of them all, *The Encyclopedia* without a qualifying adjective, was published by Diderot and D'Alembert between 1751 and 1772. This had originally begun as a project of translating Ephraim Chambers' *Cyclopedia*, an informative compendium; but it was transformed into a far more ambitious project, in which the unity of human knowledge and the power of reason was to be demonstrated in long articles. In seeing, as the contemporary authorities did, the controversial and indeed revolutionary implications of many of the articles, we are perhaps apt to forget that much space was in fact given to well-illustrated descriptions of techniques; on printing, for instance, it is a standard source for the period. Again, the example of an encyclopedia that was controversial and looked forward has not been generally followed; one wants from such a publication facts and not opinions, or so it seems, and articles have as a rule the dogmatic flavour of unrevised history and obsolescent science.

In our own century, encyclopedias seem to be aimed at those who do not bother to buy real books and do not have access to them in libraries, and one expects them to be used by children mugging up for examinations or projects rather than by scholars. Here again France presents us with another exception: the vast *Encyclopedie Methodique* which appeared in some two hundred volumes between 1782 and 1832, thus spanning an amazing amount of political change, and which contained most important publications in systematic natural history. In Britain in the early nineteenth century, *Rees' Cyclopedia* had articles by very eminent contributors and is a useful compilation although there was little original in the way it was ordered; and of the *Encyclopedia Britannica* in the second half of the nineteenth century the same can be said, for the introductory essays giving an overview of the various branches of knowledge disappeared after the eighth edition of 1860.

In the *Metropolitana*, the later volumes are a curious mixture of the encyclopedia and the dictionary; giving definitions and also, as Johnson had done, etymologies and quotations to show usage. During the nineteenth century great dictionaries of various languages were compiled, the *New English Dictionary on Historical Principles* (better known as the *Oxford English Dictionary*) being one of the greatest. It appeared slowly from 1884 to 1928, and is a monument to its time and to its editor, James Murray. Makers of a dictionary are doing the tricky task of imposing order upon flux, for the meaning and use of words changes with time and place; and the copious use of quotations from various kinds of publications can illustrate the process, for the lexicographer is not a legislator but someone aware of language changing. The arrangement of the entries so as to be clear to the reader who has to find his way around is very important, but the order of the words is given by the alphabet.

A quite different kind of dictionary is Roget's *Thesaurus*, which first appeared in 1852 and is still very valuable in later revised editions to those solving crossword puzzles, or more seriously feeling for the right word for a speech or something written. As Roget put it, 'the purpose of an ordinary dictionary is simply to explain the meaning of the words... The object aimed at in the present undertaking is exactly the converse of this: namely, – The idea being given, to find the word, or words, by which that idea may be most fitly and aptly expressed.' Words are therefore arranged in clusters under headings like 'Symmetry', 'Teaching', and 'Innocence', with an index at the back so that one can go from a given word to synonyms.

Roget had been working on his system for almost half a century, while he made a reputation for himself as a physician and man of science, lecturing in medical schools in Manchester and in London, and becoming a Fellow and then Secretary of the Royal Society. He contributed to the *Encyclopedia Metropolitana*, and wrote a treatise in two volumes in 1834 to demonstrate the existence and attributes of God from animal and vegetable physiology. It is perhaps surprising that he should have made his lasting reputation with a book about words; but logic and language are a part of all sciences, and the man of science in the early nineteenth century had to write for a general audience and could not be semi-literate. The study of language was indeed, during the first half of the nineteenth century, becoming more and more important. Language had been taken in the eighteenth century as the defining character of mankind and as the major expression of his possession of reason; but there was not then very much interest in languages themselves.

Just as it was generally assumed that the first chapters of *Genesis* described in a straightforward way the creation of the world and its inhabitants, so the story of the building of the Tower of Babel up towards heaven, and its destruction accompanied by the curse of diversity of tongues to keep mankind in place, was taken for granted as gospel truth. Thus Engelbert Kaempfer, whose *History of Japan* was published in 1727, was perplexed at the origin of the Japanese, and concluded that they were not descended from the Chinese because their language and customs were so different. He concluded that they must have come from Babylon; and that the purity of their tongue (that is, its freedom from foreign terms) indicated that they must have made their journey speedily, because otherwise they would have borrowed words from other nations as they passed through. The same framework was taken for granted by William Borlase in discussing the ancient languages and inhabitants of Britain, and especially the resemblances between the Druid and the Persian religions.

Borlase's discussion of the Druids was a part of his work on the county of Cornwall, which occupied two folio volumes; one devoted to natural history and the other to local history. At the back of this volume he added a vocabulary of the Cornish language, which at the middle of the eighteenth century was becoming extinct as the last speakers of it died. Like Hebrew, the Cornish language has been revived in recent decades, and there are now once again speakers of it and songs being composed in it. This can be done because the affinities of the language were known – it is related to Breton, Welsh and Gaelic – and Borlase's vocabulary and other such sources can be used to provide the words to fit into the structures thus deduced.

Just as Borlase was collecting his vocabularies in Cornwall, which must have been a task made easier because his informants would also have spoken English, British sailors were opening up the Pacific. Captain Cook's first voyage to the South Sea began in 1768, and he and his companions on all his voyages collected vocabularies, and tried with more or less success to learn languages at the various lands they visited. They were astonished to find that interpreters taken on board at Tahiti could be understood in New Zealand, and that the same language was also used on Hawaii; but that on Tasmania, in Australia, and in the Melanesian islands quite different languages were spoken, differing not only in vocabulary but also in structure.

By the end of the eighteenth century, the enormous diversity of the languages of North America was also being gradually revealed as the various 'nations' were visited; their languages too fell into groups, so that some seemed to be dialects, while others had something in common, and others were utterly distinct. At first it was the words of these languages that were recorded in vocabulary lists; and much attention was devoted to cases where the word for 'crab' or 'fire' or something seemed rather like the same word in Hebrew, Welsh or French. A difficulty was that those recording often differed over what they had heard; the recording of tonal differences and glottal closures is not easy with European alphabets, and adult ears are used to hearing some noises and ignoring others so that an Englishman and a Frenchman hear and record the words of exotic languages differently – as for example with the terminal vowels in Malagasy. Anyone accustomed to Cockney similarly finds the glottal clicks of Navaho easier to hear and imitate than do many Americans.

Vocabularies are only a part of a language, as we all know if we have ever tried to write French or German, or even to translate them, with nothing but a dictionary; or if we have ever received a letter written this way. And yet French and German are very close to English in structure and vocabulary; the problems are much greater when one is faced by the complications of the verb 'to go' in Athabascan languages, or the inclusive and exclusive 'we' of Malay languages. It is these characteristics, and not the coincidences of vocabulary, that establish relationships between languages, just as it is homologies in structure and not external similarities that establish classes in biology. It is therefore necessary to understand the grammar and syntax of the language, and not enough to collect words – only if, as with

Cornish, one knows the affinities of the language already can one get anywhere with a vocabulary alone.

This was something that was becoming clear in the last decades of the eighteenth century; and the great name in this connection is that of Sir William Jones. Jones had been an outstanding classical scholar at Oxford, and had been tutor in the family of Earl Spencer; he became acquainted with a number of European and Asiatic languages, and first made his name with a translation of a life of Nadir Shah from the Persian. He followed this with other translations from Oriental languages; but since one could not live in the 1770s by such work, he took to the bar and became a distinguished lawyer. It became his ambition to be appointed to one of the judgeships at Calcutta, where the East India Company now found itself administering much of a sub-continent and where Jones knew that he would be able to learn at first hand about Indian culture and also to build up a fortune. His opposition to slavery, and his support for the Americans in the War of Independence, blocked his way to this office for a time, but the influence of his friends and patrons got him the post in 1783.

At Calcutta he founded the Asiatic Society of Bengal which became one of the world's leading learned societies, and he embarked on a codification of the existing bodies of law in India. In the course of this he learned Sanscrit, and saw its close relationship to Greek and Latin; this was the more surprising because Hebrew and Arabic, the Oriental languages which (with Chinese and Japanese) were previously known to European scholars, did not show any such similarity, although Persian did. Jones died in 1794, but the study of Sanskrit was by then well-launched and with it an appreciation that languages came in some sort of families.

This period was one in which anthropology was beginning to emerge as a science, with Blumenbach at Göttingen forming his celebrated collection of skulls from different peoples and classifying them according to their dimensions. With this went the ultimately less-respectable science of phrenology, in which the character of individuals could be read from the bumps on their heads. Alexander von Humboldt in his travels in South America collected skulls; and from measurements on skulls, and also of dimensions of living men, set off by reports of 'giants' in Patagonia, physical anthropology may be said to have taken its rise. Then as later, there were some like Jones with great respect for the culture they studied, and others who found, as they had expected, that the exotic languages were impoverished and the cranial capacity of their speakers usually less than that of Europeans.

One of those with whom the study of languages and the almost zoological study of the races of men went together was J.C.Prichard, who wrote two famous books on the *Natural History* and the *Physical History* of Man. Prichard was a doctor in Bristol, coming from a Quaker family, and was an authority on mental illness as well as an anthropologist. He was a resolute opponent of slavery and the slave trade, and an object of much of his work was to establish the unity of mankind as one species: for to use members of one's own species as slaves would be wrong; but to domesticate other kinds of animals was recognised as right, so that if black men belonged to a distinct species then keeping them as slaves would be merely like keeping cows or sheep. Prichard went so far as to suggest that Adam and Eve might have been black, and that whiteness was a result of civilisation. He believed that 'the most absurd hypothesis, that the Negro is the connecting link between the white man and the ape, took its rise from the arbitrary classification of Linnaeus', and that if everybody was aware of modern and better classifications which put men and monkeys in different orders, then 'the most fanciful mind' would not have looked for an intermediate link.

Prichard used linguistic evidence as well as physical, and indeed had himself demonstrated that the Celtic languages belonged to the Indo-European group that included Sanskrit and Latin; but his general approach was that of a biologist rather than a linguist. In his writings Prichard used reports of travellers, including those of his older contemporary Alexander von Humboldt; but it was Alexander's older brother Wilhelm who was to succeed in making the study of languages central to the study of mankind. Wilhelm never shared Alexander's love for the sciences, his interests being in political theory, history, philosophy, and literature; he had a successful diplomatic career, most of it spent in Rome, and then returned to Prussia to reorganise the educational system there. The University of Berlin which he planned now bears his name; and it formed the model upon which universities all around the world were founded or refounded in the nineteenth century, with its emphasis on both teaching and research, and its ideals of *Bildung*, or self-realisation, and *Wissenschaft*, or knowledge seen as a whole.

Humboldt had hoped as a young man to write an historical work on the eighteenth century showing how the different nations had developed differently; and he had also hoped to write an ambitious and encyclopedic anthropological work. Neither of these got very far; but by the first decades of the nineteenth century he came to see that the study of languages, which had always been a source of great pleasure to him, might be the key to studying the diversity and unity of mankind. He learned Basque; and while he had, like many of his generation in Germany, an enormous respect for the ancient Greeks as the highest of civilisations, he now found that he delighted in the Basque language and culture, which were both thoroughly unlike the Greek. To him the idea that there should be one world language would have been outrageous, for the different languages like the different nations all had their parts to play and made up a diverse whole. This may seem a romantic idea, not surprising at this date; but in the same way one would never find a real natural historian wishing that there were less different kinds of creatures, though he might sometimes say so as a wry joke.

The eminence of the two Humboldts, and Wilhelm's place in the educational history of the nineteenth century as well as the learning shown in his linguistic works, meant that the systematic study of languages made rapid strides, especially in Germany. Whereas the natural historian had seemed to be classifying species that did not change at all, or at any rate had not done so since the ancient Egyptian civilisation, the linguist was faced with rapid flux. The languages spoken in Europe in the nineteenth century were manifestly very different from those that had been in use even a few hundred years ago, and quite distinct from Latin. Gradual change was of the essence in the study of languages, even when they were written down and much more when they were not. Languages also fell into groups, as natural historians had put animals and plants. In most of Europe, the languages belonged to one great family, the Indo-European; within that group there were other groups such as the Romance languages, some of which like Latin were dead while others like French were alive, and the Germanic languages. Some languages, like Icelandic, seemed to have remained fixed over centuries in which English had changed so that the older form was incomprehensible to those now living.

Whereas in Lamarck's time it was no more than an extravagent-seeming hypothesis that creatures might change slowly from one kind into another, by the middle of the century there was a model of such change; the analogy between species and languages might be more or less helpful, but at least in languages one found change of a gradual kind, at different rates in different examples, and the extinction of languages that had in the past been dominant. The Tower of Babel had disappeared from linguistics, and it could not be doubted that languages which belonged to the same family had a common ancestry; and that the closer they were, the more recent was their common ancestor. In the English-speaking world these achievements of what was mostly German scholarship were presented in the 1850s by Max Müller, first in lectures at Oxford, and then in a course at the Royal Institution in London, which were published in 1861 as *The Science of Language*.

This science was at once based upon classification, and evolutionary. Müller's enthusiasms come through in such phrases as 'It is a real pleasure to read a Turkish grammar . . . The ingenious manner in which the numerous grammatical forms are brought out, the regularity which pervades the system of declension and conjugation, the transparency and intelligibility of the whole structure, must strike all who have a sense of that wonderful power of the human mind which has displayed itself in language.' Müller saw in language the struggle for life going on at various levels, and especially carried on among synonymous words as one comes to displace the others. Müller tried to go beyond empirical fact-collection and classification, which he believed to be the early stages of all sciences, to explanation; but evolutionary explanation is not easy, for one is generally faced from the past with various written languages and others for which there is little direct evidence, and it is hard to decide for example when or how Latin gave way to Italian.

For Müller, that was not what happened; 'languages' are ideal constructs, useful for the purposes of grammatical analysis, but which should not be invested with historical reality. It was not languages but dialects that were spoken; some of the German dialects became the feeders of the literary languages of Britain, Holland and Germany, and others became extinct. To find common ancestors, one usually had to go back further than one would have expected; Italian and Latin are descended from different dialects and not one from the other. For Müller, the development of languages was controlled by two processes, called 'Dialectic regeneration', in which the language acquires new vigour from dialect usages and vocabularies, and 'Phonetic decay', as words get shortened or made easier to say in the course of usage. The change in languages was therefore subject to law, and was not simply the effect of chance; languages were not in complete flux, but evolved in certain ways. The old dogmas, that all languages took their origin from the confusion of tongues at Babel, or that Hebrew was the language of Adam and that all others were corruptions of it, were disposed of by the end of the eighteenth century, and the way was clear for an evolutionary science of language.

Müller's lectures, when published, raised him in the public estimation to the position of the leading linguist in England, and his views on language were widely accepted though naturally modified by others over the years. He was also very interested, with his early work on Sanskrit, in Indian religion, and became a pioneer in the new field of comparative religion in Britain, which brought him into contact with some of the leading men in the liberal broad-church wing of the Church of England, such as Dean Stanley of Westminster. The study of languages and their changes, revealing the great differences between the families, seemed to point to a long history for mankind – much longer than the nearly six thousand years computed from the Old Testament, though never explicitely stated there – and also cast light on the relationships and migrations of the various races of men. The term 'Aryan', which came to acquire such a nasty flavour later, was prominent in writings on linguistics, meaning simply an Indo-European group of languages and those who spoke them. Whereas languages had seemed to be distinct species, the emphasis upon dialects showed that they had a history and an ancestry; in just the same way, the old certainties about species and mere varieties in natural history were to crumble after 1859.

Evolutionary accounts of languages remind us that the borrowing was not all one way when Darwin's theory came out, though certainly the prestige his theory acquired after a few years did mean that it was applied in fields remote from natural history. There was a sense in which the idea of development was in the air, as something arising from the new idea of history that lay behind the work of men like Wilhelm von Humboldt. Whereas Agassiz and Owen had tried to find some philosophical explanation of the order of nature, showing that it could not be otherwise; Darwin, like the linguists, adopted an historical explanation. History is unlike logic or physics in that one does not expect to find a single cause of a phenomenon, or laws that would have led anybody to predict it; rather one finds a number of conditions and causes, of laws or generalisations. In the same way, Darwin's theory allowed not only for natural selection as the cause of evolution, but required the existence of variation, and also involved as mechanisms, sexual selection, and even the inheritance of acquired characters, and other agencies. In history or in Darwinian biology, a simple explanation in terms of one cause is almost certainly inadequate. It is to his theory and its effects upon ideas of order that we must now turn.

7

Development

O NE MIGHT have expected that an evolutionary outlook would have transformed the classification of plants and animals, but the effect seems to have been less than revolutionary. Features of classifications that one might have thought of as Darwinian are to be found in authors writing long before 1859, and in opponents of evolution. Thus in Agassiz's magnificent work on fossil fish, published between 1833 and 1843, he proposed a new key to the groups, based upon the scales. He divided fish into four groups, the cycloids, the ctenoids, the ganoids, and the placoids on this basis, which he recommended not only because it put like with like but also because fossil and living fish could be placed in the same system. The first two groups included three quarters of the species of modern fish, but in the earliest times there had only been the latter two. He drew a diagram, of the kind that one now sees illustrating evolution in natural history museums, where a series of horizontal lines separated the various geological epochs with the oldest at the bottom, and vertical patterns indicated the families, being wide when they are abundant in species and narrow when they are rare. For the ganoids and the placoids the patterns are not dissimilar, starting narrow at the bottom and then showing big bulges and various families running side by side; and then in more recent times, families becoming narrower and dying out so that now only remnants survive. By contrast, the cycloids and ctenoids start recently, with a history less than half as long as the others, and rapidly spread out and diverge into their present position of dominance.

Agassiz and others saw species and higher groups as perhaps having their lives like individuals, and saw higher groups diverging and contracting through time. But these successive changes did not for him establish either any progress generally through time, or any genetic connection between extinct and living species. Only among the vertebrate group did he consider that

156

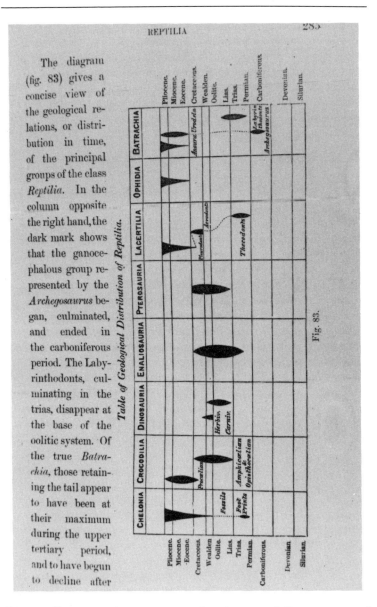

The diagram (fig. 83) gives a concise view of the geological relations, or distribution in time, of the principal groups of the class *Reptilia*. In the column opposite the right hand, the dark mark shows that the ganocephalous group represented by the *Archegosaurus* began, culminated, and ended in the carboniferous period. The Labyrinthodonts, culminating in the trias, disappear at the base of the oolitic system. Of the true *Batrachia*, those retaining the tail appear to have been at their maximum during the upper tertiary period, and to have begun to decline after

Diagram illustrating the appearance, abundance, and disappearance of reptile groups through the geological epochs. It is interesting that Darwin's most prominent opponents, Owen and Agassiz, who saw natural classification as an expression of God's mind, should have favoured such tables. From R. Owen, *Palaeontology*, 1860, p.285.

gradual and progressive development leading up to man had been indicated by recent work in palaeontology; in the other three great groups embracing the invertebrates, there was no such change, and progressive development could not be a law of nature. The vertebrates' history simply underlined the recent appearance of man; and it was generally supposed that the late appearance of the mammals generally, and of man in particular, was because they were not created until the Earth was ready for them.

It was not even certain to all investigators that there was progressive development through the vertebrates. To Lyell, with his belief in the uniformity of the processes shaping the Earth's surface through time, it was implausible to suggest that the animals and plants inhabiting this unchanging globe should have changed radically; and he was reluctant to accept both the Ice Age inferred by Agassiz, and also the late appearance of large groups of creatures. This he believed to be an illusion which would be dispelled when much more geological work had been done, when mammals would be found in old rocks – as indeed happened in the slate at Stonesfield, but never in the oldest strata. To Hugh Miller in Scotland, a man of firm religious faith as Lyell was not, it was not even evident that later fish were more advanced than older ones; it was not a matter of early fish being clumsy prototypes poorly adjusted to the world in comparison to their sophisticated successors, but simply of different features of their organisation being more complex. His book, *Footprints of the Creator*, 1849, had as its subtitle *The Asterolepis of Stromness*, the name of a formidable fossil fish on which the argument was based, and the American edition of 1850 had a preface by Agassiz.

Whether one believed in progress, in uniformity, or that each species was a new design, made fairly little difference to how one actually classified creatures. Darwin's idea that animals and plants in the same groups were related to one another again made little difference in the nineteenth century, because there were no independent ways of telling whether animals had a relatively recent common ancestor or not. The Siberian wolf and the Tasmanian wolf may or may not be put in the same group, on the basis of their various characters; if one is an evolutionist, one then infers whether or not they are closely related as an explanation of the classification, and if one is not one might still show a tree, a pattern of circles, or a diagram like that of Agassiz without feeling the need to explain it further. Darwin's theory might be a guide in the weighting of characters, like any other theory; but Darwin and his contemporaries had been warned by some of their predecessors popularising Lamarckian theories of evolution to avoid speculative family trees. Thus in the anonymous and spectacularly successful *Vestiges of the Natural History of Creation*, 1844, which sold more copies in the nineteenth century than the *Origin*, the passage from ducks to rats via the duck-billed platypus was sketched out. Had it been taken seriously it would have led to a rather different system of classification from those current; but sensible and prudent men of science preferred to work empirically, deriving family trees from natural systems and not the other way about.

In Agassiz's diagram, the vertical patterns swelled and tapered smoothly and the horizontal lines indicating the geological periods merely formed a grid. Although Agassiz was a follower of Cuvier and a catastrophist, he did not therefore believe that at the end of each geological epoch there had been a wiping clean of the slate, and a completely fresh creation. The classification of geological epochs was like the splitting of human history into different periods. Historians differ over when the Renaissance or the Romantic period began, whether it started about the same time everywhere, and whether individual artists, scholars or statesmen belong to it. A cluster of characters is all that one can find for such a period; but one might find some keys for rapid diagnosis, like domes and columns for Renaissance architecture. For historians to quarrel over whether a building or a book belongs to the Renaissance or not is to argue about terminology rather than about historical events, although at least those living then might have described themselves as witnessing the Renaissance, which is more than can be said of those living in the 'inter-war period' of the 1920s and '30s.

From 1815 the identification of geological periods from the fossils found in the strata had become the most active part of geology, especially in Britain, where Sedgwick, Buckland and Murchison were prominent in identifying 'systems' at home and abroad and naming them after the region where they found them as Devonian or Permian for example. Such systems were characterised by their fossils, but it was soon recognised that there was not a complete break and that some creatures, or families of them, went on through several systems. Statistical methods had to be used, for example by Lyell in separating into periods the Tertiary rocks of Italy where he determined the proportions of living and extinct kinds of shellfish in these geologically recent formations. There was therefore a strong element of convention and convenience about the geological systems, just as there is about the Renaissance, and yet the different floras and faunas of the periods meant that there were significant differences as well as continuities.

Just as there are arguments among historians about the applying of labels, and some historians who try to eliminate a label or two, so among geologists there were battles over the epochs. The most famous of these was over the oldest, where in Wales Sedgwick (accompanied on one of his tours by the young Darwin) and Murchison had named systems as the Cambrian and the Silurian respectively. These systems were promulgated on the basis of different fossil populations in the usual way, with certain key groups which are characteristic and lead to speedier diagnosis. Sedgwick was slow in publishing, and Murchison succeeded in taking over his Cambrian period and redescribing it as 'Lower Silurian'. To Sedgwick and his friends (among whom Murchison had previously been numbered) this seemed the basest kind of treachery and aggression; while at the time the energetic, aristocratic and powerfully-placed Murchison seemed to have won, the Cambrian did in fact survive into modern geological books. Academic rows were a feature of Victorian Britain, and bodies like the British Association with its large public debates brought them out into the open more, perhaps, than happens today; but so dry and dusty-seeming a topic as the classification of geological periods was sufficiently theory-loaded to provoke them.

The diagrams of Agassiz are the kind of thing that could be used in history, showing the areas controlled by various states for example. Thus France would have waxed and waned but survived, while Burgundy and the Venetian Republic would have become extinct. What one does not find among fossil fish (or among peoples rather than states) is the revival of a state like Poland which had been extinct for over a century; in natural history extinction is final. In 1830 this was not certain; Lyell believed that, under the right conditions, the iguanadon might make a come-back, and de la Beche drew a cartoon to illustrate the return of the dinosaurs. The life-sciences had become imbued with historical thinking by the middle of the nineteenth century, but this made surprisingly little difference in classifying; but of course people do classify societies with very little knowledge of their histories too. If species or higher groups have their history, like men and nations, it becomes reasonable to ask about origins; and indeed the question had been asked a number of times before Darwin tried to answer it in his book.

Lyell in his *Principles of Geology*, 1830, had suggested that some of the new species described each year by botanists and zoologists really were new species, that had not existed before. In his steady-state world, the extinction of species and the creation of new ones was constantly going on, and catastrophes or eruptions of the creative power were unnecessary. The effects of volcanoes and floods could be assessed in examining his views, but his idea about creation was untestable unless somebody had been fortunate enough to witness the appearance of a new species. In *Vestiges* a jerky kind of evolutionary mechanism based on embryological studies by von Baer was proposed; the development of an embryo was seen as a kind of progress down a main line, with branches off to the various groups of creatures. The branch lines were limited in number, and a duck staying on too long and taking the next turning would be born as a platypus; one of its descendants doing the same would be born a rat. One did not need to look, therefore, for missing links; and the process was automatic over time, Australia being for the author of *Vestiges* behind the rest of the world in its fauna because it was a younger continent.

The authorship of *Vestiges* was a well-kept secret, and not until after his death in the 1880s was it disclosed that the book had been written by Robert Chambers, a publisher of encyclopedias and textbooks without first-hand knowledge of the sciences, but with a capacity to see the wood rather than the trees which experts like T.H.Huxley (who attacked the book for inaccuracies) lacked. Chambers indeed admired the Quinary System because it gave an overview of the whole natural world. A.R.Wallace, who pondered on the problem, was no mere compiler, but had spent long periods in South America and in Malaya and the Indonesian islands; and in 1855 he published a paper on the appearance of new species. Geographical distribution had become his special interest, and he realised that new species seemed to come into being in regions where very similar species had been living. This had not been quite unknown before; Cuvier had recognised Jefferson's *Megalonyx* as a giant sloth, a characteristic American group, Darwin had seen the affinities between fossil and living species in

AWFUL CHANGES.

MAN FOUND ONLY IN A FOSSIL STATE.—REAPPEARANCE OF ICHTHYOSAURI.

A Lecture.—"You will at once perceive," continued PROFESSOR ICHTHYOSAURUS, "that the skull before us belonged to some of the lower order of animals; the teeth are very insignificant, the power of the jaws trifling, and altogether it seems wonderful how the creature could have procured food."

A sketch by the eminent geologist Henry de la Beche in 1830 to illustrate Lyell's notion that under the right conditions dinosaurs might return, and to question man's high position in the scheme of nature. From F. Buckland, *Curiosities of Natural History*, 1858, frontispiece.

South America, and Owen had identified fossil marsupials from Australia. But Wallace made more of it than had his predecessors, and went on to identify the line which separates the Australian fauna from the Asian and which runs between the islands of Bali and Lombok separated by a deep but narrow strait. In 1858 he sent a brief paper to Darwin with his suggestion that new species were indeed coming into being as time went on, through the effect of variation and natural selection on existing species.

Darwin's and Wallace's papers were duly presented together by Lyell and Hooker to the Linnean Society in 1858, where very little notice was taken of them. The President of the Linnean Society was Thomas Bell, who had written the volume on reptiles of *The Zoology of the Voyage of HMS Beagle* under Darwin's editorship, and was from its foundation in 1843 until 1859 President of the Ray Society, which published Darwin's studies on barnacles. In his study of *British stalk-eyed crustacea* (notably crabs) of 1856, he had had this to say about homologies:

> The typical structure of any group being given, the different habits of its component species or minor groups are provided for, not by the creation of new organs or the destruction of others, but by the modification, in form, structure, or place, or organs typically belonging to the group.

He could of course suggest no mechanism or cause that might have brought about such a state of affairs; and no doubt for him, as for most contemporaries this was simply a fact, justifying certain methods of teaching from 'typical' animals and certain ways of classifying. His use of the word 'habit' should not lead us to suppose that he was referring to behaviour; the term in natural history meant 'mode of growth'. It is nevertheless curious that Bell, and those like him who were concerned with homologies, did not realise the importance of the papers of Darwin and Wallace, Bell even remarking at the end of the year that the standard of papers presented had been high, though of course there had not been any that would really change tne course of the science.

Darwin had been working on his ideas for more than twenty years, intending to write a treatise in three volumes, but was persuaded by his friends, when Wallace's paper came out of the blue, to produce a one-volume abstract, which duly became the *Origin of*

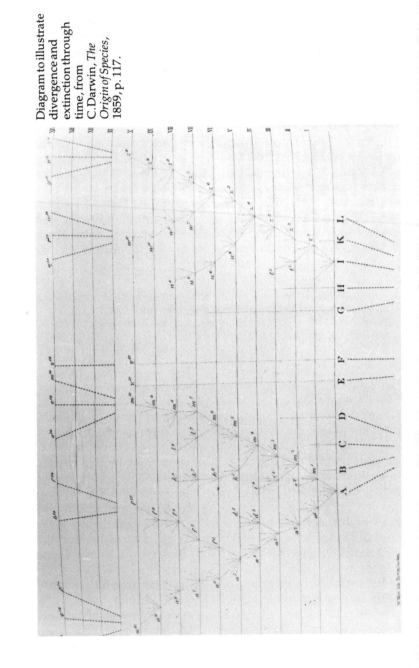

Diagram to illustrate divergence and extinction through time, from C. Darwin, *The Origin of Species*, 1859, p. 117.

SCALE OF ANIMAL KINGDOM (The numbers indicate orders:)	ORDER OF ANIMALS IN
RADIATA (1, 2, 3, 4, 5)	Infusoria
	Polyparia
	Crinoidea
	(Crustacea)
MOLLUSCA, (6, 7, 8, 9, 10, 11)	Conchifera
	Cephalopoda
ARTICULATA { Annelida (12, 13, 14)	Annelida
Crustacea (15—20)	
Arachnida & Insecta (21—31) }	Crustaceous Fishes
Pisces (32, 33, 34, 35, 36)	True Fishes
Reptilia (37, 38, 39, 40)	Piscine Saurians (ichthyosaurus, &c.)
	Pterodactyles
	Crocodiles
	Tortoises
	Batrachians
Aves (41, 42, 43, 44, 45, 46)	Birds
VERTEBRATA — Mammalia — 47 Cetacea	Bones of a cetaceous animal
48 Ruminantia	Bones of a marsupial
49 Pachydermata	Pachydermata (tapirs, &c.)
50 Edentata	
51 Rodentia	Rodentia (dormouse, squirrel, &c.)
52 Marsupialia	Marsupialia (racoon, opossum, &c.)
53 Amphibia	
54 Digitigrada	Digitigrada (genetta, fox, wolf, &c.)
55 Plantigrada	Plantigrada (bear)
56 Insectivora	
57 Cheiroptera	Edentata (sloths, &c.)
58 Quadrumana	Ruminantia (oxen, deer, &c.)
59 Bimana	Quadrumana
	Bimana (man)

ASCENDING SERIES OF ROCKS.	FŒTAL HUMAN BRAIN RESEMBLES, IN
1 Gneiss and Mica Slate system	
2 Clay Slate and Graauwacké system	1st month, that of an avertebrated animal;
3 Silurian system	
4 Old Red Sandstone	
5 Carboniferous formation	2nd month, that of a fish;
6 New Red Sandstone	3rd month, that of a turtle;
7 Oolite	4th month, that of a bird;
8 Cretaceous formation	
9 Lower Eocene	5th month, that of a rodent; / 6th month, that of a ruminant;
10 Miocene	7th month, that of a digitigrade animal;
11 Pliocene	
12 Superficial deposits	8th month, that of the quadrumana; / 9th month, attains full human character.

Table to demonstrate evolution, indicating how the natural classification, the fossil record, and embryology all point the same way. From [R. Chambers], *Vestiges of the Natural History of Creation*, 3rd ed., 1845, pp. 234–5.

Species. It is one of the few major works of science to be (deceptively) open to the layman, being in effect a popularisation of ideas not yet formally written up for men of science; it is curious that the first and second laws of thermodynamics were also published (by Helmholtz and by Sadi Carnot respectively) in popular form. The book had no notes, references, or bibliography; Lamarck's *Philosophie Zoologique* which had appeared half a century before had a much more academic look, but after all Lamarck was a professor and Darwin a country gentleman.

To those engaged in classification and description, a hypothesis about the origin of species might be interesting but was only doubtfully scientific. In the widely-prevalent view, science was either Euclidean deduction or the collection and arrangement of facts; chemists still generally distrusted the atomic theory as going beyond the facts of elements and their arrangements and combinations. Only in optics, where the wave theory of light was by the 1850s generally accepted because it seemed to account for all the phenomena, was an explanatory theory somewhat similar to Darwin's fully a part of established science. It was not fair of Darwinians writing with hindsight to deplore attacks upon the *Origin* by those who were not professional biologists, for the book was to all appearances addressed not to experts but to those interested in working out a new world-view with some basis in science. It looked indeed like an updated *Vestiges*, avoiding some of the hazards into which that book had fallen and for which it had been savaged by the learned, and with a respected name on the title page.

Vestiges had begun with a vision of a world evolving; a mass of gas rotating and forming the solar system, and then an animated globule coming into being by natural means and being the ancestor of all living beings on the Earth. By 1844, the 'nebular hypothesis' of the formation of the solar system was out of favour among those really up-to-date; and Chambers had also relied upon a report of an experiment whereby a spider had been spontaneously generated by an electric current acting on inorganic matter – a report that did not arouse much conviction among the men of science of the time. But *Vestiges* was really explosive when it urged an animal ancestry for man. The first human embryo had stayed on the main line of development longer than its parents had; and we might expect that some of our descendents would do likewise, and turn out to form a new and higher species less akin to ape. Man's free choice was held to mark him out as an intellectual and spiritual being; but by 1844 there were sufficient statistics to show that mankind was surprisingly predictable in the mass, even if individuals appeared to behave in unexpected ways. The book contains references to God, but simply as the First Cause who had set the process in motion and not the personal God of Christianity.

It was fortunate for the establishment that in *Vestiges* immorality seemed to be combined with innacuracy, so that the book could be unequivocally denounced. This had its usual effect of increasing sales, and making sure that the idea of evolution, of a progressive, inevitable and Lamarckian kind, was in the air. Darwin's book was in fact a far more serious and profound work than Chambers', although this would only slowly be realised as scientists, came to take it seriously and use its ideas, even if only at first as a working hypothesis. One problem was to convince them that any theory was necessary at all; Owen and Agassiz were very happy with a pattern of classification which was not genetic but simply indicated how creatures were in fact grouped. Their schemes, like the quinarian circles or Linnaeus' tables, could be interpreted, if one liked, as revealing the Divine plan, just as Darwin's could. What the Darwinians, like anybody trying to proclaim a new world-view, had to do was to persuade their peers that an evolutionary interpretation opened new and interesting questions, fitted existing knowledge into a satisfying and intelligible pattern, and was not merely a collection of just-so-stories.

In the *Origin*, Darwin was very cautious to keep man in the background, with the cryptic remark in the conclusion: 'In the distant future I see open fields for far more important researches. Psychology will be based on a new foundation, that of the necessary acquirement of each mental power and capacity by gradation. Light will be thrown on the origin of man and his history.' It was not until the 1870s, with the *Descent of Man* and the *Expression of the Emotions in Men and Animals* that Darwin came out into the open on these questions; but readers of the *Origin*, knowing *Vestiges* and also the attempted refutation of Lamarck in Lyell's *Principles of Geology*, were not taken in by this modest aside and realised just what was involved.

Lyell had not seriously threatened the unified world picture of the majority of men of science and clergymen, despite his insistence that the processes acting now must be seen as responsible for all past changes. This meant that Noah's Flood had to be seen as a major natural disaster perhaps, but not a worldwide catastrophe because such things did not happen nowadays; but worries about the literal accuracy of *Genesis* were not serious in the 1830s, and Lyell was appointed to a chair at King's College London, intended to be a Church counterblast to the secular University College in Gower Street. Lyell had been all right because he had insisted upon the recent appearance of man, and because of his attack on the doctrines of Lamarck, so that mankind was seen in his writings as separate from the rest of nature. It is an irony that Lyell should have expected to find mammals in the oldest strata because of his belief in a steady-state world, and yet was satisfied with man as a newcomer; but, although he had been one of those to whom Darwin had sent sketches of his theory, it was not until he had been convinced of the antiquity of man in the early 1860s that he came to accept evolution by natural selection.

The threat in Darwin's writing was sharply perceived by Owen whose Platonic and orderly scheme of things was in strong contrast to Darwin's, and by Wilberforce who saw man's uniqueness and moral capacity being denied. He also, like other reviewers of the book, remarked that it was presented as no more than a series of probabilities and was not beyond doubt as science ought to be. Both of them emphasised man and his place in nature; the curious thing was that Owen had been one of the very few experts to give a welcome to *Vestiges*, though a qualified one. His dislike of the *Origin* seems to have been partly because of Darwin's emphasis on slow and continuous change without sharp discontinuities between different species, and partly because of his jealousy – he was a man of great ability and eminence but nobody who had to work with him enjoyed it.

When the British Association met in Oxford in the summer of 1860 it was clear that there were going to be clashes about Darwin's theory. Darwin was ill and stayed away. In the mid-nineteenth century, the British Association was a spectacle rather than a sober forum for the announcement of new experimental observations, though some of the sectional meetings were pretty dry. It met in a different town each year, and the drumming up of public interest in science as an intellectual activity and as a potential source of wealth was a major point of the whole enterprise. People were attracted to the meetings to see the lions perform, in the days before one could watch them in television programmes. As on television it is the persons of restless and versatile intellect who come over better than the experts who carefully weigh every word, so it was in the more exciting sessions of the British Association; where (as in scientific lectures at the Royal Institution in London) most of the audience had come to be entertained as much as informed.

Exactly what happened at the meeting is surprisingly unclear, for it has become surrounded in myth and most accounts resemble the story of St George and the Dragon rather than an actual historical event. A recent investigation of the three apparently independent accounts of the clash, in the *Lives* of Darwin, Huxley, and Hooker published at intervals from twenty to fifty years after the event, has shown that they were all in fact cooked up by the same people and represent a synoptic tradition of the Darwinian Church Scientific, to use one of the ecclesiastical metaphors in which Huxley the agnostic delighted. The only contemporary version, which is one often quoted, is from a letter written afterwards by an undergraduate who was strongly Darwinian and appears to have added various details and phrases to increase verisimilitude in what would otherwise have been a rather plain tale of a drawn match. It does not seem that until some twenty years later there was any general feeling that it was improper for a Bishop to have been debating a topic that should have been left to experts, nor that the Bishop had lost the battle. Indeed, it seems to have been unclear who his opponent was, although the later stories always cast Huxley as the hero.

According to such sources as there are, it appears that the Bishop was witty, and that Huxley expostulated indignantly and either went a bit too far (casting himself as a soldier or a bulldog) or was not really audible or intelligible, or both: and that the Darwinian case was made out by Hooker, from Kew, the eminent and formidable botanist and friend of Darwin who had been converted some time before Huxley. Indeed to him and to his fellow-botanist Asa Gray of Harvard, Darwin's theory seemed to offer explanations of the distributions of certain plants that were otherwise inexplicable. Gray found affinities between plants of Eastern North Amerca and Japan while those of California were related to those of Central and South America. He invoked alternating periods of warmth and of glaciation so that a flora which had occupied the arctic regions in a warm epoch had moved and diverged southwards as the climate grew colder, down the eastern coasts of America and Asia. Hooker had also found the need to invoke evolution in accounting for the flora of Tasmania and other southern islands, and these two botanists had been Darwin's first two converts.

Huxley was in 1860 an eminent physiologist, but he had joined in the attacks on *Vesitges*, zestfully pointing out its blunders. In *Vestiges* and in Wilberforce he deplored the intrusion of amateurs into his science; he, with Hooker and others, formed the X-club which functioned as a pressure group of agnostic professional scientists, playing an important role in the scientific establishment in the years after 1860 and showing itself as intolerant of dilettanti, heretics or apostates as any Inquisition. The British Association meeting marked the transition in Huxley's career from the scientist to the sage, deploring amateurs in science and yet prepared to hold forth on any topic himself. This was obvious to him twenty years later, but not at the time to anybody else, and the meeting therefore took on significance as time passed.

Wilberforce was a talented man and an energetic Bishop, not altogether popular with clergy, to whom the epithet 'Soapy Sam' was later attached when he had been involved in the disciplining of some over-liberal clergymen. His father had been largely responsible for the stopping of the slave trade, and he himself had spoken at the British Association on an earlier day of the 1860 meeting against the suggestion that Africans were not ready for trade or missions, his record on racism being much better than that of many scientists of his day, including Huxley. It was with the status of man that he was most concerned; and then as now, when biologists try writing about man, especially in works apparently aimed at a large public and with a catchy title, they have to expect to be criticised by others who may feel that mankind is equally their province – indeed, we probably all feel that.

It is therefore perhaps surprising that Wilberforce did not cite theologians or philosophers as his authorities, but eminent men of science like Owen; his own science was naturally out of date, and no doubt based upon Paley, Bridgewater Treatises, and semi-popular works rather than monographs translated from the German. He may well have felt that anybody who had read mathematics could judge any other scientific literature; but he was not putting himself forward as a Bishop whose duty it was to pour anathemas upon unchristian or heretical doctrines. Rather he was acting as the spokesman, at a time when two cultures had not yet emerged in England, of the general intellectual and scientific community faced with the revival of a superficially convincing but specious doctrine, which if accepted would have unpleasant social consequences. Even Huxley admitted that if a 'General Council of the Church Scientific' – what we might call an international scientific conference – had been called in 1860 it would have condemned the Darwinian theory. We have seen in the twentieth century enough consequences of scientists feeling themselves immune from the criticisms of outsiders, to feel that those who think they see defects and dangers in a piece of scientific reasoning should not be debarred from saying so.

It is to Draper, whose speech preceded that of Wilberforce and bored the audience, that we owe the idea that nineteenth-century intellectual history is chiefly a matter of warfare between science and religion; military metaphors were popular at a time when many in Britain were seriously alarmed about the intentions of Napoleon III, and an invasion even seemed possible. The picture may fit some parts of the last decades of the nineteenth century, but it does not fit 1860; one need not admire Wilberforce to see that the odium seems to have been on the other side, in that minority of Darwinian agnostics who felt themselves beleaguered by their peers, and who later had their revenge when their doctrine had prevailed.

It is always wise to outlive one's opponents and get the last word, as Huxley did with Owen, contributing a chapter to the official *Life* of his old enemy. The confrontation between Wilberforce and Huxley is, when divested of myth, reduced to not much more than intellectual entertainment; but at the same meeting there was a clash between Own and Huxley that was of more scientific significance. It directly concerned man. Owen declared that man was quite separate from the apes, and should be in a separate order, the *Bimana*, quite distinct from the *Quadrumana* where the apes and monkeys were to be found. This was for the first half of the nineteenth century the orthodox view, although it had not been for Linnaeus; but authors were happy when they could find new evidence to confirm this belief, and Owen considered that he had done so in dissections of the brains of apes and men.

Despite the interest in man's place in nature, it is curious how little work had been done on the animals which were his nearest neighbours, even if only to see how far away they were. The various anthropoid apes were not clearly distinguished, the term 'orang-outang' being frequently used to cover all of them; all that was generally accepted was that man, the rational animal of

171

Aristotle or the talking animal of the eighteenth century, was different from them. Tyson's monograph on a chimpanzee written at the end of the seventeenth century was still a standard work. In 1862, in order to justify his remarks about man in 1860, Huxley published his little book, *Man's Place in Nature*, which was intended to establish that man's place was among the apes. It was written as a popular work, and proved to be so; once again the argument was being taken to the general public by the Darwinians rather than being settled in scientific journals and societies, although Huxley and his associates were intent upon the professionalising of science. The book contains three chapters, the first on the natural history of the apes, the second on their anatomical relations to man, and the third on fossil remains of man.

The first chapter is historical and discursive. Much of it is 'scissors and paste' stuff, culled from early works and quoted directly; but Huxley was becoming a stylist, and was thorough, and the work was competently done. There had recently been some excitement about the gorilla, a previously barely-known species, of which accounts had been given by an explorer, Du Chaillu, which had been denounced as sensationalism by many eminent zoologists; he reported that they went into the attack drumming their chests. In fact, much of what he said about them seems to have been true, but there has always been a strong tendency for tales of sexual and other aggression to be attached to what are really a rather shy group of animals. For some reason, Huxley appended to this chapter a note on cannibalism in Africa, in a manner which would have struck those with a deeper concern for the natives of that continent as flippant.

The second chapter is the one that is of interest to us because it is concerned with the relationships between the apes and with man, and was based upon work actually done by Huxley and his associates. Beginning with references to 'time-honoured theories and strongly-rooted prejudices', Huxley went on to present his argument 'in a form intelligible to those who possess no special acquaintance with anatomical science'. There follows an exposition of embryology, to demonstrate that man originates by physical processes identical to those of other animals; and then the reader is urged to consider whether or not man belongs in the same order as the apes 'with as much judicial calmness as if the question related to a new Opossum'. What Huxley did was then to show that the differences between the apes themselves and between them and man were just those which elsewhere in zoology characterised different families and genera but not orders. He did an especial comparison between a man and a gorilla, paying particular attention to the skull and other parts of the skeleton – notably the hands and feet, which had been supposed particularly human – and then to the brain. He found that men differed more from each other in cranial capacity than they did from apes, although the smallest skull was that of 'an idiotic female' rather than a healthy adult; and that whatever characters were taken the differences between the apes were similar to their differences from men.

Since for Huxley, and for most of his contemporaries, taxonomic groups higher than species were artificial anyway – for most followed Gray here, rather than Agassiz for whom all groups were natural – it was only a matter of convention and not a matter of substance whether one regarded the differences of men and apes as a matter of orders, families, or genera; but Huxley was going rather far if tidy-mindedness was all that he had in mind. Today, naturalists do debate these questions; within a big group such as the herons, different authorities will draw the boundaries differently, but will not publish a popular and controversial work to justify themselves. Huxley's object was to urge that if man were separated from the animals by no barrier higher than separates them from one another, then 'if any process of physical causation can be discovered by which the genera and families of ordinary animals have been produced, that process of causation is amply sufficient to account for the origin of Man.' He believed that Darwin had indeed described such a process. This argument may be right, and would nowadays be generally accepted, but it did not recommend itself to such Darwinians as Wallace and Asa Gray. It is like Galileo's, that because the Earth resembles the Moon and the planets in certain respects, it must move as they do; where again the conclusion is generally accepted, but the reasoning is not watertight.

In this second chapter, Huxley had shown that for consistency one should not put man into a distinct order, though he did remain in a distinct family in which he was the only genus and species. He was careful to point out that this still made him very different from the apes, and that there were no missing links between him and them, or indeed between the various kinds of apes. Some of his remarks could certainly be taken by the sensitive to imply that some races of men were superior to others, but men did not for Huxley grade into apes – they were still zoologically a long way apart. His third chapter was on fossil remains of man. Once one had denied that the apes were a kind of man, man's place was not really a matter of fundamental concern; but his ancestry was, and here Huxley's remarks about Darwin's theory led on to the remarks about fossil man.

When Darwin published the *Origin*, the evolutionary history of living species and higher groups was still shrouded in mist. It was not until some time later, in America, that the history of the horse was worked out, and that 'missing-link' fossils like toothed birds came to light to give some sort of direct confirmation of the theory; although of course they do not prove it true for each 'intermediate' form might be a new creation and not the descendent of its predecessor, and parent of its successor. If the history of no species was clearly established, that of man (and of the apes) was especially uncertain. No clearly-human fossils were known in the early years of the nineteenth century, and those which had been attributed to giants or to those drowned in the Flood were assigned by more expert anatomists to different species. The recent appearance of man seemed to be indicated by this negative evidence – the result we might say of looking in Europe and North America rather than in Africa – and this was consoling to those who wanted to preserve some authority for *Genesis,* and who feared that if men knew they were descended from apes they would spend their time monkeying about, which does not necessarily follow.

In 1833 there was published the account of bones found at a cave at Engis in Belgium which included some human bones apparently coeval with those of mammoths. These were the first remains of what was later called Neanderthal man to be found, and Schmerling who wrote up the discovery was concerned to know to what race (rather than species) of man he had belonged. The Belgian finds were fragmentary, but in 1857, in the Neanderthal, a complete skeleton was found in a cave. Schaaffhausen in his description of the bones referred them to 'a barbarous and savage race': to Huxley, the skull showed some resemblance to that of an Australian aborigine; its external ape-like characteristics were to be set against its brain capacity, which was about the same as that of modern men. The fossil was therefore not an intermediate between apes and men, but simply a human skull showing reversion towards its primitive stock. All that Huxley could conclude was that we should keep looking for earlier remains of *Homo sapiens* (for he evidently believed that the Neanderthal man belonged to our species):

> In still older strata do the fossilized bones of an Ape more anthropoid, or a Man more pithecoid, than any yet known await the researches of some unborn palaeontologist? Time will show. But, in the meanwhile, if any form of the doctrine of progressive development is correct, we must extend by long epochs the most liberal estimate that has yet been made of the antiquity of Man.

Huxley's book was therefore curiously tentative and theory-laden. The problem about man's ancestry was not to be solved in the nineteenth century, but the question of whether he had been around for a few thousand years or for much longer could be given a definite answer. The authoritative work here was Lyell's *Antiquity of Man*, published in the year after Huxley's book, 1863. One of the problems about human remains is the habit people have of burying the dead, so that especially if there are only a few examples, one cannot be sure that they are really contemporary with other bones in the cave or in the same deposit. Because the question of man's ancestry was connected with other beliefs about religion and society, there were also strong preconceptions leading towards credulity or scepticism when faced with human bones or reports of them; people knew what they wanted to believe. There was also the possibility of faking, by planting bones and then finding them. The most famous case of this was the Piltdown Man, a 'missing link' whose skull was composed of cleverly-stained parts from those of a man and an ape; this was found in 1913, and assigned to a new genus, *Eoanthropus*, and still duly featured among man's ancestors in an exhibition at the Festival of Britain in 1951, before new tests revealed the forgery. Every discovery contained the possibility of misinterpretation or fraud, and as with the Piltdown skull many experts might be deceived for a long time.

Lyell had long been familiar with Darwin's ideas, but his rejection of 'progressive development' was very long-standing, and his *Principles of Geology* had contained a chapter in refutation of Lamarck. Although over the years he had been sent letters and abstracts of what became the *Origin*, he did not announce himself a convert to Darwin's views until 1863, and fully in the tenth edition of his *Principles* in 1867–8. He had been for some years pondering on the problem, which for him was particularly related to man; his notebooks on the species question, which have recently been published, are fascinating because of the open-mindedness and balance with which he presented the arguments to himself. His slowness in coming to a conclusion mildly exasperated Darwin, and is in great contrast to the way Huxley worked. Lyell well knew how many fossil groups were defined on the basis of a single specimen, often very fragmentary; and that while Owen had triumphed in describing the moa on the basis of a single bone, he had also identified a fossil as from a monkey and much later had to revise this opinion, deciding that it came from a pachyderm. Palaeontological knowledge was for Lyell fragmentary and provisional, and he would not go so far as some of those who knew less about the subject.

Lyell, with his firm belief in the uniformity of processes, had been reluctant to think or publish much about man until well into the 1850s; in France the work of Boucher de Perthes at Abbeville, where he found flint tools associated with the bones of extinct animals, and in India and Britain that of Hugh Falconer, was little noticed until 1858. Falconer, a surgeon working as a geologist and botanist for the East India Company, had found in the Sewalik Hills the remains of a mammalian fauna including a tortoise six feet high and twelve feet long; and he had made himself an authority on the various extinct species of rhinoceros and elephants. He had found that this fauna contained a number of creatures like the crocodile of the Ganges which were still living in India. He found in Indian mythology stories of enormous tortoises, and began to wonder if man had coexisted with them, for he saw that while faunas had changed over time, the species composing them had not all come into being or died out simultaneously.

On returning to England in 1855, he became interested in the geologically-recent deposits there, with particular reference to man. Buckland's *Reliquiae Diluvianae* of 1823 had seemed at the time to say all that could be said about caves and their contents, neatly explained in terms of Noah's Flood; later when the Flood had disappeared from geology, caves seemed rather embarrassing, and any traces of man found there ambiguous. Most caves had been disturbed by fossil-hunters, and could not yield definite evidence. But in 1858, at Brixham in Devonshire, a new and large cave was discovered behind a quarry. Falconer called the attention of men of science to it, and secured a grant from the Royal Society to enable a scientific investigation to be carried out. Great care (of the kind that we now associate with archaeologists' digs, and which they learned from geologists) was taken to plot the position and depth of everything found, and there could be no doubt that the human remains and artefacts must have been contemporary with the extinct animals also found there.

Lyell had been a member of the committee set up by Falconer and his associates to supervise the dig, and in 1859 he told the British Association about it; in the following year, at the celebrated Oxford Meeting, Murchison announced his belief in the antiquity of man. As Falconer put it in a fragment, 'Primeval Man and his Contemporiaries' unpublished at his death in 1865: 'The public mind, led in its convictions by the favoured few whom it elected to follow, was craving for accessible information on the subject; and a well-considered digest of the evidence of the case, in all its bearings, was the only thing wanting to secure for it the accepted belief of mankind'. Falconer seems to have intended to write such a work himself, and the fragment marked the beginning of it; but Lyell, well-qualified by his legal training to digest evidence, got there before him. In Falconer's view, his book, written by one who had not been concerned in the actual work done, made it all look too easy, and was not fully just to those who had painfully worked their way towards the new knowledge.

Now as then, men of science do poach on one another's territory, and the names best known to the general public are not always those who stand highest in the esteem of their colleagues, being often past their best work. Falconer and his associates (notably Joseph Prestwich) had not had much to do with the classification of man, for they all took it that the human remains belonged to our species, but they had done careful classification of the animals of geologically recent time, and had had made the later 'Tertiary' and the 'Quaternary' epochs of central interest in geology. Lyell was not just a populariser basking in the limelight, however, even though he could not match the palaeontological skill of Falconer or the stratigraphical technique of Prestwich; he had pondered about species and the succession of faunas and floras, and ever since 1830 had been revising his *Principles* and *Elements* in successive editions to take into account his own researches and those of others, build them into his synthesis, and present them to a public much wider than one would have expected for a scientific monograph or textbook, for they do not quite fit these modern categories. But his notebooks are more interesting than his *Antiquity of Man* because he was working painfully rather than digesting smoothly.

The need for caution was in fact shown in 1863, the year of the publication of Lyell's book and thus of great interest in 'pre-Adamites'. Human remains were found at Moulin-Quignon near Abbeville where Boucher de Perthes had been working for many years, finding worked flints but no bones or teeth of men among those of extinct animals. A jawbone and a tooth were discovered by one of the workmen; having long ignored these researches, Parisian scientists now became interested and members of the Academy of Sciences were prepared to vouch for the genuineness of the finds. Falconer and others from Britain were unhappy, and believed that the human remains had been planted by one of the workmen, who had apparently also begun a brisk trade in flints both ancient and modern. In the event, after a conference both sides struck to their guns; but a very dubious find never became a Piltdown scandal, partly because of the publicly expressed doubts, and partly because the bones were no kind of 'missing link'. It is always prudent on these occasions to blame workmen rather than colleagues.

Lyell in his notebooks distinguished firmly, as Huxley writing of 'progressive development' did not, between 'out and out pro-gressionists' and Darwin; and he recognised firmly how loaded discussions of species must be: 'The ordinary naturalist is not sufficiently aware that when dogmatizing on what species are, he is grappling with the whole question of the organic world & its connection with the time past & with Man; that it involves the question of Man & his relation to the brutes, of instinct, intelligence & reason, of Creation, transmutation & progressive improvement or development.' If this were strictly true, then nobody would ever dare to name a new kind of fly or centipede, and the passage reveals Lyell the philosopher rather than the working geologist; but we can admit that the idea of a species is loaded with a great deal of metaphysics, and that classification necessarily has a theoretical background.

It is curious that on the same page he has a note to the effect that Darwin said that the difference between the case of the three species of anthropoid apes and the various races of man was that among the apes there were no intermediate gradations: 'Had there been no links, the Negro & White Man would have been made species'. This was written in 1857; what is curious is that, even if Darwin had said it, it was not true. If what is meant is that the orang-outang, the chimpanzee and the gorilla are not inter-fertile, whereas all the races of men are, then this was what was normally meant by a species or a race and the two cases were clearly different. If men of different colours did not interbreed, then by definition they would belong to different species, but the point is entirely hypothetical. If on the other hand, one used the other definition of species, based on similar anatomy, then the races of men are much closer to one another (as Huxley was to show in 1862) than are the various species of apes – we are brothers under the skin, even if skull shapes do vary. The theoretical background that lay behind a good deal of the scientific discussion of man was that some races were inferior to others; and because of the way things were, being pale, long-nosed, hairy, competitive, and raucous of voice went with superiority.

Lyell used comparisons with languages to clarify his ideas of the emergence of species; and he was also rather pleased with the thought that the creation of man might not have been very different from the birth of a genius to parents of ordinary capacities. He believed that astronomy provided an analogy for the classification of something in flux, with its constellations, which in 'ten millions of years' will all have vanished because of movement of the whole solar system and the 'fixed' stars. This was a time scale comparable to that of species:

> But the fact of the instability in the forms of the constellations which are varying constantly, annually, hourly, does not prevent them from remaining for thousands, nay for hundreds of thousands of years, so unchanged that the Navigator may overlook such trifling modifications . . . appearances which last for thousands of generations of Man are to them at least realities – they partake of the fixed and the immutable.

Elsewhere he compared species with periods in geology, both alike in being abstractions from a continuum but making geology a science, and presumably Darwinian botany and zoology also.

Lyell was, after 1855, gradually coming to accept the implications of Darwinian science, and seeing them perhaps more clearly and dispassionately than is usual in those who have actually to work in a science. His notes show his reluctance to discard a First

Cause, although his geology had been for thirty years directed at explaining all events without recourse to divine intervention, and indicate his deep repugnance at admitting an animal ancestry for mankind, although he admitted that to trace him back to a 'clod' was no more dignified and exalted than to trace him back to a highly organised creature. He realised that the chimpanzee was not an ancestor of man, and that if we had an ape for an ancestor it was an earlier ape-like creature, *Dryopithecus*, (described in 1856), and that it had taken at least a million years for him to evolve into us. He noted that millions of men died before birth or very soon after it, and that the problem of when the merely animal foetus became a rational being with a human soul was greatly debated between theologians – as of course it still is now that pre-natal deaths are brought about by doctors in abortions. As the individual passes imperceptibly from animal to man, so perhaps had the human race in the past. The creation of new species was perhaps so slow and imperceptible as to escape detection.

Lyell's reluctance to see any law of progressive development was encouraged by his noting that there seemed to have been less development in Australia, although that continent was no younger than the others. In this he is unusual, for his contemporaries who had seen the coming of the Reform Bill, the electric telegraph, the railway and the steamship were firm believers in progress, and found it easy to believe that the law of improvement held throughout the world. Darwinian evolution, with its ups and downs as species boom and crash, may have suited certain economic perceptions of the day, but it formed a gloomier creed than most people liked. Goal-directed change rather than just change, or in effect a Lamarckian rather than a strictly Darwinian idea of evolution, was the preferred view even of some professed Darwinians like Asa Gray. In a strictly Darwinian world, the only teleology is what Huxley called the 'new teleology', which means that one can say that creatures will not have developed or retained an organ unless it is of some use to them, and that they will therefore be adapted to their station – if they were not, they would not have survived. Buffon had thought that there were some creatures like the sloth that were so ill-designed as to merit our pity; but writers of natural theology, and naturalists going outside the closet, had soon shown that the sloth in its habitat got along very well and could do without our sympathy. They put this down to God's benevolent designing of things, whereas Darwin put it

180

down to natural selection; much of his writing is not fundamentally so very different from Paley's, and belief in a world of struggle with as many disasters as triumphs is more in accordance with orthodox Christianity than is the doctrine of progress. It should not have come as a surprise to any but the most liberal Christians to hear that selfishness was natural, or that belief in Providence was a little like whistling in the dark.

In seeing racism in Victorian authors, one is often looking through the wrong kind of spectacles. It is true that even those who were keenest to emancipate the slaves and evangelise the Africans often regarded them as inferior beings, who would need hundreds of years to catch up with Europe or North America. One point was that it was only among men and domestic animals that races had been seriously studied, and among animals it had been more a practical than a theoretical study. The investigation of small differences among bones or teeth (with men it was skulls and their cranial capacities that were especially studied) was also a technique that had been developed for distinguishing species rather than races, and differences tended thereby to acquire specific rank. But the most important point was that those living in the technically advanced countries, going through a period of political and economic revolution, could not but be conscious of the gap which separated them from the 'naked savages' of Tierra del Fuego or Australia. The prejudice was as much against the 'backward' as it was against specific races; and it had long been noted, usually with some complacency, that such peoples seemed to be inexorably exterminated (directly or indirectly, through rum and smallpox) before the advance of superior peoples.

This was something that could be given a kind of Darwinian explanation which removed it from the sphere of morality, and it produced economy of thought because one could lump all 'backward' peoples together as we do in talking about the 'third world'. Those who forgot that any 'missing link' would have lived over a million years ago noticed that primitive peoples, who were no longer usually seen as noble savages, lived not very differently from social animals. The rather more sophisticated, with the historical feeling that was new in the nineteenth century, recognised in savages their own ancestors who had painted themselves blue and danced and grimaced in an attempt to frighten the Romans. With the interest in history went the belief that its stages were necessary – something we still see in Marxism. Just as an embryo and then a child has to go through stages in growth, so peoples have to grow up; and if the process took Englishmen two thousand years, it will no doubt take Fuegians just as long. This kind of historicism reinforced the belief that embryological development was a recapitulation of racial history, and the scientific doctrine then reinforced the historicism.

The feeling about other races was then as much that they were behind Europeans, at least of the professional classes, as biologically distinct from them; but in an evolutionary world, to be ahead was to be superior, and perhaps the distinction made little difference. Certainly the brotherhood of man seems to have become harder to feel in the nineteenth century: the first British rulers of India in the late eighteenth and early nineteenth centuries seem to have shown a mixture of firmness, rapacity and sympathy not unlike other rulers of that and other countries, but their successors displayed an ineffable superiority often accompanied by a generalised benevolence, treating heir subjects more like children. Their predecessors had negotiated with local rulers on equal terms, and had fought battles against formidable generals, and could hardly doubt that Indians could administer or command. No doubt Arnold's pernicious revival of the public schools (creating, in alliance with industry, 'two nations' at home) had something to do with this earnest or effortless superiority, but that cannot be the whole story because other European nations behaved not very differently. Men were classified into civilised and barbarian, rather as they had been in Antiquity, both at home and abroad; but Darwinian theory could give a biological turn that was new and nasty.

Once, like Lyell, people had accepted that species must change so slowly that for practical purposes one could go on treating them as stable, like constellations, then Darwin's theory made comparatively little difference to taxonomists. One still went by multiple criteria, giving more weight to internal rather than external characters; vestigial organs, like our appendix, had been taken as homologies, and were now taken as evidence of descent, but for practical purposes that came to the same thing. The great new task was to search out family trees; some of these like early etymologies were not very plausible, but others like that of the horse seemed well worked out. Palaeontologists had looked for characteristic fossils to determine to what period a formation should be assigned; now they looked through the strata in order to follow development. In 1856 *Dryopithecus*, who might be an ancestor of man, had been found; and in 1891 Lyell's prediction that in equatorial Africa or in the East Indian Archipelago one might hope to meet with 'lost types' of the anthropoid primates was fulfilled with the discovery of the skull of Java Man, *Pithecanthropus erectus* (the erect ape-man), named in accordance with Ernst Haeckel's prediction. By the end of the century, an evolutionary history and a kinship with the apes was generally accepted for man.

Other classifications, whether of chemical elements or political systems, were also given an evolutionary explanation in the last decades of the nineteenth century, and any classifying was generally done in order to work out a genealogy. Historical explanation was everywhere, and the point Huxley may have made in 1860 that origin and worth are not connected was often ignored. In the opening years of the twentieth century, the Darwinian idea of slow or almost imperceptible change was for a time abandoned in favour of a more jerky kind of evolution, depending upon grosser mutations, in which (as in *Vestiges*) the distinctions between species were no longer blurred but remained definite, but the pattern in which classification was no more than a way of indicating history remained in biology as elsewhere. That is nowadays only one impulse behind taxonomy; and in our epilogue we must continue the story of classification up to our time.

8

Epilogue

I N THE CENTURY since Darwin died, his views have been subjected to all sorts of criticisms but an evolutionary perspective has been the only one that really offered any kind of satisfactory explanation of the variety of animals and plants. While there has been argument over mechanisms of evolution, and while the whole subject of genetics has come into being, to doubt evolution would mark one as eccentric or heretical. Those who do, usually content themselves with pointing out phenomena so far unaccounted for but cannot present an acceptable alternative synthesis. Subject to the caveat that all science, if not all human knowledge, is provisional, one could say that evolution is as well-established as anything else we take for granted.

The success of Darwninism has duly meant that classifications, both in the life sciences and elsewhere, have been given an historical explanation. But this has not meant that classification is obsolete. In biology, the notion of species has lost some of the precision that Darwin's older contemporaries like Agassiz had given it. Accepted species may interbreed occasionally, and there is an area of uncertainty around many groups which is revealed in the use of terms such as superspecies and subspecies. The vision of Agassiz and others of species and higher groups having a lifespan, flourishing and dwindling over a period, becomes more complicated if part of a population undergoes rapid evolutionary change into something rather different, while another part, in different surroundings, is becoming rare and eventually extinct. There are not the sharp lines between groups, or the definite dates at which species came into being or died out, for which pre-Darwinians sought. Plants and animals still have to be named and knowledge of them systematised so that it can be easily found and passed on; and for these purposes the Linnean binomial nomenclature still seems to work surprisingly well. Taxonomists still, as they did in Linnaeus' day, argue over the wisdom of revising genera, and hence names, too painstakingly, and over the question of splitting or lumping; but they do so in the knowledge that they are classifying something in flux.

Darwinian explanations of how given species came into being still have the look of more or less likely stories, and taxonomists still have to rely upon characters rather than genealogies. In the classification of chemical elements, contemporaries of Darwin gave evolutionary explanations of the Periodic Table. The most notable of these was perhaps William Crookes for whom the various elements in their families (envisaged by him in a three-dimensional array) were manifestations of the swings of some cosmic pendulu, and the Rare Earth metals (or Lanthanides) found only in Sweden were a kind of living fossil group which had never become fully differentiated and turned into distinct species. The swinging pendulum is closer to the kind of occult progressive power of nature that lay behind Lamarck's vision, rather than to the natural selection (a metaphor hard to apply to chemical elements) of Darwin; and it was far from the kind of mechanism that people as hard-headed as chemists would be expected to enthuse about, even if it was put forward by a man very distinguished for his discovery of Thallium with the spectroscope, and for his work on cathode rays.

Nevertheless, speculations like Crookes' inclined chemists to accept that, like the species of the life sciences, their elements were not permanent but might change one into another. The various families like the alkali metals (the most familiar of which are sodium and potassium) looked a bit like real biological families, or genera; and terms like affinity had long been used of chemical elements though, as in pre-Darwinian biology, it was a metaphor not implying anything corresponding to blood-relationship. Eminent chemists like Davy and Faraday in the first half of the nineteenth century had cautiously expressed the conviction that the alchemists might have been not too far from the truth in their conviction that one element might be transformed into another; just as it struck zoologists as odd that God should have independently created so many kinds of creatures with very small differences, so it seemed to Davy incompatible with the simplicty and harmony of nature that the metals sodium and potassium which he had isolated should be so similar and yet irreducibly different. William Prout had gone so far as to propose that all the elements were polymers of hydrogen, or some lighter substance; and his hypothesis,

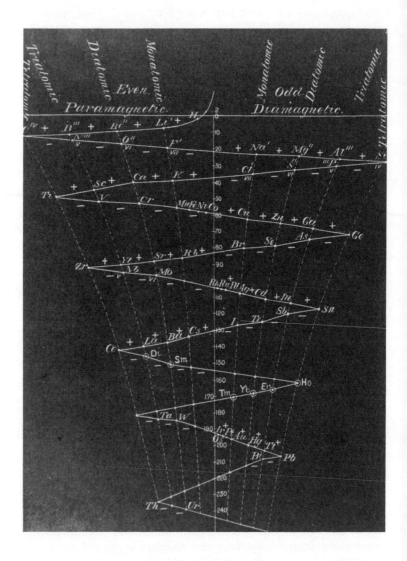

Periodic Table set out to show a hypothetical evolutionary mechanism of pendulum-swings, by William Crookes, from *Report of the British Association*, 1886, p.567. The vertical; scale is of atomic weights. The 'odd' and 'even' elements differ in chemical and magnetic properties in a way that could not be explained before the work of Rutherford and Bohr.

falsified, it seemed, when the atomic weights turned out not to be exact multiples of that of hydrogen, survived in amended forms through the century to tantalise investigators. Many atomic weights were nearly multiples of that of hydrogen, far more than would be expected in a chance distribution; and as more and more elements were discovered, caesium and rubidium being exceedingly like potassium for example, those who looked for simplicity in nature and in science found that they could not believe, like Dalton, that the elements were composed of distinct atoms which had remained unchanged since the creation.

In 1868 Norman Lockyer found some lines in the spectrum of the Sun which did not seem to belong to any element known on earth, and he suggested that this was a new element, calling it helium. Nothing was known of its properties, but in an evolutionary time it could be supposed that it was the primeval element from which all others had come. Lockyer was also to apply evolutionary ideas to the classifications of stars, where he saw two distinct sequences; following on the work of Secchi whose classification appeared in English in 1868. Lockyer's helium was in 1895 identified on Earth, as a gas occluded in certain minerals, by William Ramsay who was bringing to light a whole family of elements, the inert or 'noble' gases. He found it to be denser than hydrogen, and therefore not the parent of all the elements. Also in 1895, Rutherford came from New Zealand to Cambridge, to work under J.J. Thomson. In 1897 Thomson succeeded in measuring the ratio of mass to charge of the 'corpuscles' which made up the cathode rays of Crookes, seemingly establishing that this radiation was corpuscular rather than simply a wave motion as light was supposed to be.

Thomson, in choosing the name 'corpuscle', was the heir of Newton and Boyle, for whom corpuscles were the basic building blocks of the world and were all composed of the same stuff. Thomson wrote that the explanation for the cathode rays being identical, whatever the cathode be made of or whatever gas be in the tube, was 'that the atoms of the different chemical elements are different aggregations of atoms of the same kind'. The corpuscles were rechristened 'electrons' and recognised as only one of the constituents of all matter over the next ten years; but physicists like Thomson were prepared to consider the possibility that atoms might be split, or split up spontaneously, largely because the chemists' classification of elements existed and had been given an evolutionary twist.

It was Rutherford in the opening years of the twentieth century who worked out formally a theory of 'sub-atomic chemical change' to explain what happened when radioactive elements decayed; and this theory, despite the logical difficulties of admitting that an atom (the word means something that cannot be divided) could be split, soon prevailed. Rutherford's collaborator Soddy came to London in 1903 to work with Ramsay, and they followed the first laboratory transmutation to become a part of public scientific knowledge when they observed the radium emanation 'radon' spontaneously disintegrating and generating helium and other elements, the helium being detected with the spectroscope.

Rutherford then worked out a general atomic model of a nucleus with electrons in orbits; and in the second decade of the twentieth century this model was applied by Bohr in detail to account for the arrangement of the chemical elements in the Periodic Table. The nuclei were supposed to be positively charged, each element being characterised by its charge; the negatively-charged electrons were in orbits strictly determined by the quantum theory, the orbits being in definite sets or 'shells'. When one set was filled, the next then began to be occupied, those corresponding to the lowest energy being first filled in the normal state of the atom. Elements like sodium and potassium were very similar because they both had a single electron in their outermost shell, but potassium had one more shell filled than sodium. Some complicated families like the 'rare earths' which had puzzled Crookes were supposed to represent the complete filling of an inner shell. The theory was not merely a just-so story or working hypothesis, for it accounted exactly for the spectrum of hydrogen (the simplest element) and took into account experiments on the scattering of α-particles. Thus the classification of elements was explained, and this might be seen as a branch of science passing from stamp-collecting to physics.

Things are not as simple as that, however. To apply the theory to atoms much larger than that of hydrogen involves making various assumptions as well as most calculations, and one cannot yet predict chemical properties from electronic orbits, or orbitals as they are now called. When chemical properties have been determined, then one can work out electronic configurations; but it cannot be done the other way round. The Periodic Table has not become obsolete, being replaced by a few equations, but still has its place on the walls of chemical lecture rooms as a convenient classification, summarising a great deal of knowledge, and allowing predictions and comparisons to be made that one would not think of without it. Although Bohr's theory seemed to have more precision than Darwin's theory of evolution, it still did not make classification obsolete.

Electrons and the other sub-atomic 'particles' which have been discovered or postulated in the twentieth century themselves posed problems to the taxonomist. To J.J.Thomson, it seemed as though they must be particles not quite unlike minute billiard balls; and his experiment of 1897 vindicated this belief. Other experiments proved what others, especially in Germany, had believed, that the cathode rays were a wave that could be diffracted like light; and upon this property the electron microscope depends. It is one thing to have a few anomalies in a classification, as in zoology where we generally have no problem in deciding if we are faced with a mammal, bird, or reptile and only occasionally meet creatures like the duck-billed platypus; but if the basic building material of the world is neither a wave nor a particle but obeys equations appropriate to both in different circumstances, then one might well feel rather unhappy. Indeed the classification of fundamental particles seems to be sufficiently unorganised and baffling to make the physicist envy the entomologist: who has many more species to deal with, but has a clear definition of his field, and had, by the twentieth century, usually established which apparently different specimens are really male and female, or adult and juvenile, forms, while the transformations of particles remain ambiguous and uncertain. Indeed, in his new book Gary Zukav presents physics, not as a science of particles, but as a cosmic dance, in which energy manifests itself in different ways. Such a dynamic science requires a new taxonomy; and once again we find entities in flux to classify.

Darwin's theory did not revolutionise taxonomy in the life sciences although it did explain how the classes might have come about and did lead to the working out of some family trees, such as that of the horse. What has happened in taxonomy since Darwin is generally that new characters have been taken into account, and that evolutionary theory has sometimes provided a guide in the weighting of characters. The external features on which Werner, in his mineralogy, and Swainson, in his zoology, had relied have been steadily less credited, but this was something that had begun well before 1859. In birds, the form and coiling of the intestines for example is taken as a better guide than the shape of the bill, which may be similar in species living a similar kind of life but not closely related. Cuvier and Darwin made biology an historical science; and, as in history, new interpretations arise when old evidence is seen in a new light, or a new source or kind of source is discovered, so classifications are revised.

The reinterpretation of evidence may be illustrated from studies on the shape of the skull, which led to the placing of the birds of paradise near the starlings. At the turn of the eighteenth century, Goethe had been pleased to discover in the human skull traces of the intermaxillary bone which is prominent in some other animals, thus providing more evidence of the unity of creation. Goethe did not go on to suppose a common ancestry, but his reasoning was otherwise not unlike that of later taxonomists. His contemporary, T.I.M.Forster, published a work on swallows in 1808 which had reached a sixth edition by 1817, and used the form of the skull for different purpose. He was attracted like many able contemporaries by the science of phrenology, founded at this time by Gall and Spurzheim, in which character was indicated by bumps on the head, which themselves indicated particular development of parts of the brain. Forster found that the bump of wandering was prominent in the swallow, and that therefore it must migrate, as Collinson had thought and Linnaeus denied.

Phrenology did not live up to its promise, but the episode illustrates how a new science may cast light on an old one. By the middle of the nineteenth century, the classification of minerals was based on a mixture of chemical and 'external' features. In biological classification, the new taxonomy of the twentieth century made great use of biochemistry. Discrimination of egg-white proteins, for example, can be used to classify birds, and the distribution patterns of amino-acids can be used to classify plants. Perhaps because, even among biologists, one still finds a belief in a 'mechanical' hierarchy of sciences in which physics and then chemistry seem the most fundamental, chemical evidence has often been given great weight although a classification based upon it may conflict with one taking into account a wide range of characters. One has to be careful too, because for example caffeine is found in various plants, including those from which we get tea, coffee, and cocoa, which cannot be closely related; and it must, like many external characters, have arisen through parallel evolution. Genetic evidence, from counting of chromosomes to study of DNA,

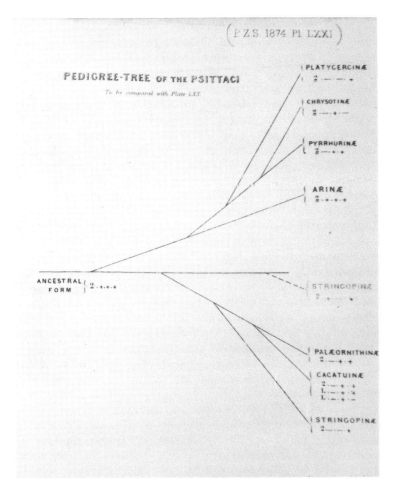

An evolutionary tree for an actual group of creatures, the parrots, by a Darwinian, 1874, based largely on the form of the skull. From A.H.Garrod, *Collected Scientific Papers*, ed. W.A.Forbes, 1881, p.247.

again cannot (so far) be divorced from attention to other characters. A natural classification still cannot be made dependent upon a single character, however fundamental we may believe it to be; though single characters may be very useful diagnostic keys or guides, like 'spot-tests' in chemical analysis, after the group is

191

defined in terms of multiple characters.

The difficulty of weighting characters, and the degree of judgement inevitable in such activity, has led to an arousal of interest in numerical taxonomy in which each character is assigned equal weight. This idea goes back to Adanson, the contemporary of Linnaeus; the fundamental problem about it is the same as that facing 'Baconian' science in which one collects facts without any preconceptions. Since the number of facts one comes across, or the number of characters an organism possesses, is infinite (or strictly, indefinite), it is necessary to have some idea of which are relevant. Actual science is always the outcome of a tension between the expected and the unexpected; and in taxonomy one is inevitably picking 'relevant' characters. There must also be many characters in plants and animals as inaccessible to us as chromosome numbers or biochemical data were to Adanson, so that a numerical taxonomy can, like any other, only be provisional.

Bacon and Baconians hoped to make science democratic, as the educated judgement akin to conoisseurship required to perceive causal connections or natural groupings gave way to a more public and mechanical way of proceeding. In numerical taxonomy, the idea is that as many characters of various organisms as possible are enumerated, and groups then formed from those which share most characters, rather as Baconians hoped to move from numerous instances to laws of nature. Here the computer has come to the aid of the taxonomist, for it can assimilate and organise the mass of data. As with anything computerised, the result depends on what has been put in, and there is a grave danger that the illusion of enormous accuracy can be given when groups are made up on the basis of the proportions of characters they have in common; the taking into account of one more character might perhaps have made a considerable numerical difference. As with induction, one hopes that in the long run one will get nearer and nearer the truth. The method so far has only found application for special problems; should its conclusions (like those of Linnaeus) ever seriously conflict with those of able and experienced taxonomists giving rather more weight to characters that seem fundamental on theoretical grounds, then it is to be expected and hoped that the mechanical and numerical system would give way to the more intuitive, but one cannot depend on it, such is the prestige of numeracy and computers.

As well as biochemistry and genetics, new developments

within natural history have furnished data for the taxonomist. In the nineteenth century, he worked in a museum as a rule, using bones, skins, or specimens preserved in alcohol or other fluids. In the field, the 'collector' who furnished him with specimens concentrated upon building up a large collection; when he saw an animal, he shot it. Binoculars were little used for the observation of living creatures; but with the establishment of zoos with scientific purposes in Paris at the end of the eighteenth century and then elsewhere, some investigations into animal behaviour were made. In the London Zoo, for example, the toucan duly demonstrated how it managed to go to sleep tucking its bill under its wing. Creatures in zoos do not behave just as they do in the wild; but animal behaviour was regarded by many as something to provide anecdotes for amateurs and not as a part of serious science. There were careful studies of natural history by people such as Charles Waterton, the traveller in South America and country squire in Yorkshire; but this was seen as being on the fringe of science, and certainly of no relevance to taxonomy.

Darwin, another traveller and country gentleman, changed this. If man was not quite distinct from all other animals, then we could expect to see in animal behaviour analogies to human behaviour, just as their organs and ours are similar. In illustration of this idea, Darwin published in 1872 his book on the *Expression of the Emotions in Man and Animals*, which was one of the earlier scientific books to be illustrated with photographs. In this book Darwin had set out to refute the idea of Sir Charles Bell the eminent neurologist that we have certain muscles in order to smile, laugh and communicate generally with our fellows. Darwin argued that similar muscles in other creatures perform much the same functions. From such studies came the idea that animal behaviour might be studied in order to cast light on human behaviour, though the modern science of ethology is a product of the twentieth century.

Perhaps more seriously, Darwin's theory meant that one ought to take animal behaviour seriously. Natural selection exerted sufficient pressure to eliminate useless activities, and so the songs, dances, roarings and grimacings of animals must have a function just as their limbs and organs do. These things were therefore worthy of serious study and were as important as comparative anatomy. Just as homologous organs indicated to the Darwinian a community of descent with modification, so advantageous behaviour might be inherited and closely related species of birds, for example, might have similar calls, mating behaviour, and nesting habits that could only be observed in living specimens in the wild. Families such as the crows and the pigeons do indeed show homologies in behaviour, just as they do in their inward parts; and naturalists can predict the behaviour of a newly-discovered species from that of its relations rather as Mendeleev could predict the properties of new chemical elements using his Periodic Table. It is partly because of the change in biological theory that the naturalist of the twentieth century will be keen on the conservation of wildlife, whereas his predecessors tended to be keen on getting a good bag and almost had to be a good shot.

Similar behaviour, like similar structures, may be found in creatures that are not closely related but inhabit similar places; and so predictions here, as elsewhere, may be wrong. A test of reconstructions and predictions came in the middle of the twentieth century when, in December 1938, a curious fish five feet long later identified as a coelacanth was caught by a trawler off East London in South Africa. This group of fish had flourished for a long time beginning about three hundred million years ago, but were thought to have been extinct for at least fifty million years. The man who identified it, J.L.B.Smith, was a chemist turning himself into an ichthyologist; and over the succeeding years he became the great authority on the fish of southern Africa (developing a numerical key to groups) as he sought for another coelacanth, because the soft parts in the first one had been thrown out as they started to decay before an expert could see them. The chance to dissect a fish of such a very ancient lineage was like a trip in a time-machine; for the coelacanth resembled astonishingly closely the reconstructions based on abundant fossil evidence, but only hard parts leave fossils. Eventually Smith became convinced that the home of the fish must be the Comores, islands off the northern part of Madagascar; and in 1952 after photographs and notices of a reward had been circulated among fishermen there, a second coelacanth was duly brought in, collected by Smith in a dramatic dash in a military plane, and thus available for study. Since then, a number of others have been caught, and islanders had apparently been finding and eating them from time to time in the past. To men of science, and indeed to the public, the discovery of a creature believed to be extinct, a 'living fossil', was a dramatic reminder of the provisional nature of science, especially since this one should have been big enough to be noticed; but it was also a vindication of palaeontology and taxonomy that it could be identified and classified so fast.

While the public was suddenly being thus introduced to ichthyology as an exciting science, weeping tears of joy with Smith when he saw the fish at the Comores, elsewhere in Africa evidence of much closer ancestors of man was being uncovered by the Leakeys. Lyell and Darwin had pointed out that if man had recent ancestors in common with the great apes, then it was likely that any fossil remains of them would be found where the apes live; that is, in Africa or in Indonesia. Early human skulls were indeed found in Java and near Pekin; but it has been in Africa that the major finds have been made, and it seems clear that that continent was the cradle of the human race. In the nineteenth century and the first half of the twentieth, zoologists made aware by Huxley and others of the close similarities between man and the apes, expected to find that they had diverged from a common stock relatively recently; and the Java man was given the generic name *Pithecanthropus*, or ape-man, because he was supposed to come soon after the divergence and to show intermediate characteristics.

Evidence from fossils and from biochemistry has now pushed the divergence back to some twelve million years ago, to *Ramapithecus* which seems to have moved out from the forests into more open country, and there evolved into two lines, *Australopithecus* and *Homo*; it now seems as though these two lines were independent, and that for a long time there were two separately developing sequences of increasingly man-like creatures until eventually *Homo* proved the more successful stock and *Australopithecus* died out. These are matters of probability and the weighing of evidence of a fragmentary kind; but it seems beyond doubt that there were, by about three million years ago, creatures in Africa that can properly be referred to the genus *Homo*, though not to our species. Man is certainly not the newcomer that he was generally supposed to be in the pre-Darwinian era, and he is much older than Lyell or Darwin had supposed.

One result of these fossil discoveries, associated with tools of an elementary kind, has been to induce a less fine taxonomy among students of mankind. Darwinism generally goes perhaps with lumping rather than splitting; but to found new genera of man-like creatures upon a fragment of cranium or a tooth, especially from a period when rapid evolutionary change was probably occurring, would be to impose an artificial grid on a fluid reality. The Java man is therefore now assigned to the genus *Homo*; whereas Neanderthal man is once again believed to belong to *Homo sapiens*, and to differ from modern men only sub-specifically, having particular adaptations to life in the Ice Age. On this view, the various races of modern man are not even of sub-specific rank, and all of us of whatever colour or head-shape belong to *Homo sapiens sapiens*. Some would prefer to omit the third name altogether here and elsewhere as leading to too fine a taxon-omy that does not reflect how breeding populations work; but at all events there is nothing significant biologically to separate living men, and little to separate us from men living as long ago as one hundred thousand years; whereas the apes are much more distinct from us than Linnaeus and the early Darwinians believed, though undoubtedly our nearest relatives.

To be akin to apes was particularly alarming to Victorians because they were generally supposed to be violent and obscene animals. In the eighteenth century, it was possible with Lord Mon-boddo to see the savage as noble, and the ape as nobler still; but when some of the crew of one of Cook's ships had been eaten in New Zealand, and a boat's crew from la Pérouse's expedition killed in the Pacific by islanders to whom they had been trying to behave well, belief in the noble savage started to wane and was replaced by the idea of the 'civilising mission' or the 'white man's burden'. If natives were savage, how much savager must apes be; and stories of the ferocity of apes, and their lusting after women, came to replace those in which the orang-outang was seen as one of nature's gentlemen. Neither story was based on careful obser-vations of the various species.

In fact apes seem to be rather peaceful animals, living mostly on a vegetable diet in quite small groups. Aristotle's classification of man as the political animal seems to be vindicated as far as separat-ing him from apes is concerned, for man is a social animal as they are not. It has been suggested that his emergence from the forests, and then his taking to a partly meat diet, brought a need for cooper-ation in larger groups, and thus helped to bring about the emerg-ence of language; for the eighteenth-century idea that language separated men from other animals seems still to have something in it even though we know so much more about animal communi-cation. There is no especial reason to suppose that early man was fierce, or that urban vandalism or warfare is the emergence of characters from apes or earlier ancestors; for hunting does not require particular ferocity, especially towards one's own species, and to talk of selfish genes is to use metaphors to a degree that makes 'natural selection' look like a sober statement of fact – even if the model helps us to understand curious features in the sex-lives of birds and (especially) bees.

It seems more reasonable to connect aggression with the later phases of human development, with larger agricultural societies, which have resources to defend, and with spare time in the autumn and winter. The Classical interpretation of the ages of Gold, Bronze, and Iron, as a story of moral decline from primitive freedom to urban civilisation and warfare, seems to have at least as much in it as the nineteenth-century view of progress from the Stone through the Bronze to the Iron Age. The trouble is illustrated, for example, in the Biblical myth of Cain and Abel, where the covetous and aggressive farmer kills his good brother the herdsman; and part of the attraction of the Wild West is its telescoping of history as the Indian hunter is driven out by the cowboy, still retaining some primitive simplicity and glamour, followed by the farmer with his fences and his cash-crops, and the miner carrying on his extractive and polluting activities.

To extrapolate back from the simplest societies of our day to the social organisation of *Homo erectus* over half a million years ago, or *Homo habilis* about three million years ago, is like using living coelacanths to cast light on those found fossil; no doubt the resemblances are very close, but neither animals nor societies are really 'living fossils' going on completely unchanged with the passage of time. And before succumbing too completely to nostalgia for the pre-agricultural period of human existence, we should remember that there are some aggressive hunting and food-gathering peoples, like certain Red Indians; and that while poetry and some works of art go happily with primitive society, there are some agreeable things, like books, that do not. Perhaps the computer and the micro-chip will eventually bring us the leisure enjoyed by our ancestors in their hunting lives, or even when they pursued subsistence agriculture in warm climates; but whether we shall be as good at filling it remains to be seen. We should remember that most of their leisure time was not taken up with the composition of epics or cave-paintings, but probably with gambling, storytelling, drinking, taking the cow for a walk, and occasionally feasting, and no doubt we can find analogues for those things.

Thus while our community of descent with the apes is undoubted, if we want to learn about ourselves the proper study of mankind is still man rather than chimpanzees; there are some things we can learn from apes about our ancestry and about what makes us distinctively human, but there is much more that is interesting to learn from one another. In the natural classification, man belongs among the primates and is not in some separate division all by himself; but the differences in classifications are as important as the resemblances between the things classified and we should not lose sight of man as the political, rational and talkative animal. In the eighteenth century we had *L'homme machine*, and in the twentieth we have had him described as the naked ape; but such catchy terms do not do more than indicate one aspect of man's place in nature.

To classify man among the apes, or any other animals, again does not imply any kind of biological determinism. There are many things like flying unaided that men cannot do for biological reasons, but this does not constitute determinism. The place to which he is assigned in nature does not however, imply that he is the victim of uncontrollable atavistic urges, or must in certain situations behave in certain ways, as termites are usually found to do. We know from introspection, from the behaviour of our family and friends, and from reports of anthropologists and historians, how varied are human responses to similar states of affairs. Fortunately we and others generally display a certain consistency, or social life would become chaotic; and in the mass, as the author of *Vestiges* and those who run public opinion polls know, men are pretty predictable. But this behaviour cannot usually be best accounted for in biological terms, by some analogy with birds defending a territory or seals fighting for a mate. We know what we mean when we describe somebody as a lion or a snake in the grass, and it is on that level of metaphor that it is wise to stop. Certainly biological determinism is no more plausible than the older Calvinistic or economic determinism; and we should be careful not to suppose that there is a hierarchy of sciences, so that a biological explanation is superior or more fundamental than one in terms of a social science. This is like supposing that biochemical evidence outweighs all other in taxonomy.

Drawing the line between men and our ape-like ancestors is a problem for the palaeontologist, but not really for anybody else because we can no longer suppose, with our nineteenth-century ancestors, that some races of mankind are closer to apes than others. It becomes of interest to all of us if we try to decide what are the characters which we would regard as distinctively human. Some of these would be structural, like an upright walk and the ability to oppose finger and thumb so as to get a firm or a delicate grip with the hand. With this would go capacity to use tools, and capacity for language; attempts to teach apes to speak have ended in failure, but they can be taught sign languages used for the deaf and dumb – speaking depends partly on intellect, and partly on the structure of the throat and mouth. Cranial capacity was made much of in the last century, for man the rational animal must have the largest brain; but it is very hard to decide just what cranial capacity separates man from a mere hominid, and cranial capacity in modern man is rather variable. The jerky mechanism of *Vestiges* in which ducks gave birth to platypuses, which in turn gave birth to rats, is as faulty in application to humans as to rodents; in a period of rapid evolutionary change, the frontier zone separating men and apes must have been imperceptibly but irrevocably crossed probably by several groups independently.

Since then man has become stabilised, and he has it in common with coelacanths that he is not changing into anything else – one could call him a degenerate species for this reason, or else an exceedingly successful one, in danger of succumbing to his own success and destroying himself. In pre-Darwinian biology, man was the summit of creation, and all other creatures were seen as steps on the way that led to his appearance; biology grew out of medicine and agriculture. To the early evolutionists, putting the Great Chain of Being into motion, man was the end to which the whole creation strove, and all other creatures were his ancestors in this progressive scheme. Cuvier's separation of the animal kingdom into four great branches meant that man only stood at the head of a part of it; but the tendency was still there to see animals as imperfect men. To Lamarckians like the author of *Vestiges*, it was indeed possible that from man a new creature no longer half-akin to ape might emerge; but at the present, man was the crown of the system. Animals were described as more or less perfect as they approached to man in organisation. In a Darwinian world, perfection has nothing to do with similarity to another species but with adaptation to an environment, and with flexibility to cope with changes in that environment. What counts is success. We have still to some extent to come to terms with this idea, and to see 'lowly' and 'nasty' organisms as just as much the product of the evolutionary process as we are, rather than as stopping places or cul-de-sacs on the road that led to us.

Whether it need affect our self-esteem that we seem to be the outcome of processes which have also generated all other organisms, and which do not seem to have been directed at the production of mankind, is open to doubt. The metaphor of natural selection seems to imply that Nature selects like a breeder of horses or cattle and has chosen us, but it is no more than a metaphor. The Darwinian world on the other hand is a coherent one, more so than its predecessor; for just as in the Copernican world, all planets (including the Earth) had the same physics, so all species have come into being in the same way rather than by a series of independent creations. Those who believe in God may easily find evolutionary creation a grander conception than endless separate creations of often exceedingly similar kinds of creatures. Natural theology since Kant has to be a matter of starting with some convictions about God, and then testing and refining them by experience of nature; to attempt to prove the existence of God with logical rigor from science is hopeless, but one may reasonably hope to discern purpose in a world of chance if one sets out looking for it, as Darwin found evidence for evolution once he had the glimmerings of the idea to account for his early observations. The great mistake would be to fall into the error of the Deists of the eighteenth century who supposed that they had the key to understanding God's mind, for our knowledge is always provisional and imperfect; and faith in God is properly more akin to faith in students or in Britain than to belief in some scientific theory or hypothetical entity.

Scientific knowledge is always provisional, and despite the great support that Darwinian theory has received over the last hundred years there is always the possibility that it might go the way of other syntheses highly effective in their time, like Daltonian atomism or Newtonian mechanics, which have had to be highly modified so that while they have their uses they can no longer be believed in. Classifications, like the 'laws' of chemistry and physics, can outlast theoretical systems; the coming of Darwinism made relatively little change to taxonomy, and should it go then natural classifications would survive its departure and presumably be better accounted for by its successor. Even 'laws of nature' are overtaken by the march of science: Boyle's Law is now only held to apply strictly to 'ideal' gases rather unlike real ones, and Bode's Law of planetary distances has been given up. In the same kind of way, classifications we suppose to be natural, or anyway more natural than their predecessors, may well turn out to be artifical, ready, like the Linnean system, to be abandoned in due course.

If we think, as they did in the early nineteenth century, not of the natural system, but of the natural method, then this prospect need worry us less; for every change is simply the more rigorous application of the method of using multiple characters, and one may reasonably hope that the method is progressive and will steadily bring an increasingly natural classification. Because more evidence is taken into account, man's place in nature is more accurately known than it was in Huxley's day, or in Linnaeus'. This is a pragmatic line, akin to the hope that science generally is a matter of progress; and that while in both theoretical science and in taxonomy there are wrong turnings or blind alleys, like the phlogiston theory or the quinary system, our understanding of the world in the long run has steadily improved since the days of Bacon, Kepler, Galileo and Descartes, the inventors of modern 'scientific method'. What should perhaps astonish us is not that at the frontiers of knowledge there are uncertainties and disputes, but that so much of it is uncontroversial, or as the physicist John Ziman calls it, 'consensual' or public knowledge.

What seems unavoidable is that this consensus rests upon assumptions or paradigms that cannot be proved. In a book of 1890, E.B.Poulton, one of Huxley's best pupils and Professor of Zoology at Oxford, wrote: 'any scientific work which I have had the opportunity of doing has been inspired by one firm purpose – the desire to support, in however small a degree, and to illustrate by new examples, those great principles which we owe to the life and writings of Charles Darwin, and especially the pre-eminent principle of Natural Selection.' One could not hope for a firmer subscription to the thirty-nine articles of Darwinism, unless it were the affirmation of A.C.Marsh at the American Association for the Advancement of Science in 1877 in Tennessee, when he said: 'I am sure I need offer here no argument for evolution; since to doubt evolution today is to doubt science, and science is only another name for truth.' Thomas Kuhn has rightly drawn attention to the role of dogma in science, where there is no room for the luxury of agnosticism. Marsh's work on the ancestry of the birds, and Poulton's on the coloration of animals, only made sense in a Darwinian paradigm, and the success of their researches reinforced their belief. Those many people in Tennessee who were not convinced by Marsh's appeal to authority were not engaged in research in biology or geology.

One field where Darwinism and classification were closely connected was the study of 'mimicry', which cast fresh light on the problems of affinity and analogy. H.W.Bates had travelled with Wallace in Brazil and had stayed on there collecting from 1848 to 1859 on the Amazon. He noticed that various species of creatures, notably but not only butterflies, which he collected and which were from genera palatable to birds and other predators, had come to resemble very closely species from genera known to be unpalatable. This 'mimicry' was a matter not only of appearance, but even of manner of flight, and would deceive not only birds but also entomologists who were not being very careful. Classification based on externals would therefore have put these creatures into quite a different group from their real or natural place based on the study of internal characters. For Bates, Darwin's idea of natural selection provided the clue, and indeed the phenomenon was 'a most beautiful proof of the theory of natural selection'. Palatable species had been modified to look like distasteful ones, because in each generation individuals looking unpalatable would have been more likely to survive. A similar process had led to distasteful or stinging creatures coming to look alike, although they too were of quite distinct kinds; for if predators have to learn by trial and error which creatures not to eat, it is to the advantage of all that they should only have to do it once rather than have to taste one of each kind of unpalatable species. In such extreme and complicated cases of adaptation to environments, the theory of descent with modification could explain the incompatibility of the internal and external characters, and provide a reason for giving little weight in classification to the latter.

To some Darwinians it seemed as though there might be another way of determining relationships, and thus achieving a natural and evolutionary system by a short cut avoiding the careful balancing of numerous characters. Embryology seemed the way, with the magic phrase 'Ontogeny recapitulates phylogeny'. What this meant was that the individual's embryological development represented a rapid run through the whole history of his race. A man began as an invertebrate, then became a fish, and then passed through other groups on his way to birth; and Haeckel in the latter years of the nineteenth century made much of this idea, using it to construct numerous family trees. In the twentieth century the notion, like an old soldier, faded away except from some popularisations: for the human embryo is never a fish, and embryological development is better seen (as by Darwin's and Huxley's hero, Karl von Baer) as a process of passage from the general to the individual. Only in a very general way, as another factor to be taken into account, can embryonic development be taken as a guide to classification and ancestry.

Haeckel's theory had implied that higher features of an organism were added on at the end of its embryological history; and it was with some surprise therefore that it was noticed that the resemblance between men and foetal apes is much closer than between the adults of the two groups. It looked almost as though man was a kind of ape that had not come to full term; or in the decent obscurity of a learned tongue, he was a paedomorph. S.J.Gould in a fascinating work which is at once history and also a contribution to theory, *Ontogeny and Phylogeny*, has separated this into two processes; a curtailment of embryological development, and a prolonging of childhood, to both of which he plausibly attributes great importance in the evolution of man and other creatures. While it allows him to look for 'bushes' rather than 'ladders' in man's ancestry, with stable branches and rapid spreadings into twigs from time to time rather than slow and steady progress, it gives no evidence on its own for natural relationships independent of natural classification.

Before 1859, the paradigms were rather different. The debates over classification, natural or artificial, and man's place in nature would be, as Dawkins suggested, not worth anything if the change in paradigm had been so profound that earlier discussions were almost in another language. Physics before Galileo or chemistry before Lavoisier is sometimes dismissed in the same kind of way; and yet if one takes a little trouble to study these disciplines, one finds that those before the 'revolution' were engaged on an enterprise not wholly different from their successors. Promising babies have indeed sometimes been thrown out with bathwater, like the study of analogies between Swainson and Bates. Through these revolutions a great deal of the science remained unchanged, although different emphases and interpretations were put upon the data; and often the new entities were a kind of mirror-image of the old ones, as Lavoisier's 'oxygen' was of 'phlogiston'. Similarly, to study France under the *ancien regime*, or Imperial China or Czarist Russia, is not only interesting in itself but helpful in understanding the country today; for about as much seems to survive from the past as it does in countries with an apparently smoother course of history, like Britain and the USA.

In the same way, Aristotle, Linnaeus, and Cuvier really were Darwin's 'gods', who had posed questions and sketched answers for him to puzzle over and improve; and much of his writing can be seen as natural theology reinterpreted, an inverted Paley. The great arguments over the status of man had occupied the century before the *Origin of Species* came out; while today we know much more about man's ancestry that Darwin and his contemporaries did, work on the fossil record is not likely to tell us very much about ourselves that is really interesting and important, and that was not known to politicians, philosophers, novelists, historians and indeed to any reflective person long before Darwin was born. Ziman's remark about experimental psychology applies also to biological reasoning about man: 'we must ask whether we do, in fact, obtain from this research some truly invaluable insights that we would rather trust than many alternative sources of knowledge about the ways of our fellows.' There would be little gain if mankind were emancipated from the old priesthood only to be made subject to the dogmas of the new priesthood of scientists; for the old ones at least were supposed to be primarily concerned with individual man's happiness and ultimate fate, and with morality, whereas scientism is a poor sort of creed. From science and history together we may hope to see the dogmas for what they are, necessary but limited, and work out a critical and personal view in which the faith of our fathers may well also feature.

Morality could be said to have entered natural history with the realisation that it is not much use ordering the world if we also destroy all or much of that order – if the owl of Minerva really presides only over life that has ended, then scientific knowledge is bought at too high a price. This is something which has come into science in the twentieth century, and in practical terms it may turn out to be as important as any of the intellectual revolutions in the history of science. Those of us with six children should not take a Darwinian satisfaction in our biological success (if one ever had time for that), or so it could be argued, but should feel guilty at contributing to the overpopulation of the globe. Like most global arguments, this has various sides to it; a decline in population in Britain, for example, which closes colleges and schools and accentuates the move towards an ageing population is not good for anybody except geriatrics specialists, and vets, since those who forego children seem to keep dogs instead! Not everybody can become a professional taxonomist or historian of science; arguments of the kind, 'if everyone did it where would we be?', do not apply to many human activities. And if, as seems plausible, those from large, cheerful and improvident families grow up happier than only children of whom too much is expected, then it might be that fewer people having larger families would be much more satisfactory than trying to make everybody have one or two children. It seems a pity, anyway, that so much trouble goes into stopping life at its inception, and into extending it towards its close; it is a curious feature of mankind that it is those with least to live for who most fear death.

Concern with the environment reminds us that, unlike the animals, we can foresee and plan, adjusting our lives not in accordance with territorial imperatives but with cultural and political norms. These are not imposed upon us, but are the result of the interaction of societies and nature; and science is much the same. We order the world in accordance with certain preconceptions, and with a logic that goes back at least to Aristotle: we have no guarantee that our classifications, or any empirical knowledge, are reliable and that we are not involved in a delusive game or a conspiracy; but under criticism and faced with new observations, our systems are refined, and such knowledge is the best we can have. We can feel with Davy that it would be dull if one could be sure of science; otherwise we might share the gloom of Lagrange that there was an order in the heavens, and Newton had found it out already just leaving tidying-up for his successors.

All our thought, therefore, is provisional, even if it is abstract and deductive, for it is then of uncertain application to the world. Alexander von Humboldt's *Cosmos* was a personal view at a particular time; and all our science is like that. His title was from the Greek word meaning 'order': our word comes from the arranging of the threads in weaving, reminding us that ordering is an activity and not something we can perceive passively – not that we perceive anything passively, for perception depends on selection and classification. Philosophers of science have on the whole said too little about this fundamental preoccupation, and it is agreeable to find Ziman writing: 'Indeed, in all branches of the natural sciences, whether or not they are amenable to mathematical analysis, the problem of *classification* is fundamental, and cannot be settled by an arbitrary convention which has no roots in underlying reality.' In ordinary life and in the sciences we are for ever classifying and ordering, beginning with crude bifurcations but working towards a natural system in which everything must have its place, and travelling hopefully even if we know we shall never arrive, for to mankind there are no final solutions. For special purposes we remain content with the crude divisions, into friends and foes, or into game and vermin; but such artificial systems do not satisfy the reason. Although we can never be certain of it, in ordering the world we take it for granted that there is a real order there.

Suggested Further Reading

1. For Thomas Kuhn's view of science, see his *Structure of Scientific Revolutions*, 2nd ed., Chicago, 1970. R.Dawkins, *The Selfish Gene*, Oxford, 1976, is an example of a model (of blood being thicker than water) taken rather more seriously than more sophisticated authors would allow, and of metaphor out of control, but is nonetheless good reading. On the Cambridge network of the nineteenth century, see S.F.Cannon, *Science in Culture*, New York, 1978. For Chemical theory and classification, see my *Transcendental Part of Chemistry*, Folkestone, 1978; M.P.Crosland, *Historical Studies in the Language of Chemistry*, 2nd ed., New York, 1978. On classifying in different societies, M.Douglas, *Purity and Danger*, London, 1966 and *Natural Symbols*, London, 1970; E.Durkheim and M.Mauss, *Primitive Classification*, ed. and tr. R.Needham, 2nd ed. 1969. On psychical research, see A.Gauld, *The Founders of Psychical Research*, London, 1968.
2. T.S.Kuhn, *The Copernican Revolution*, Cambridge, Mass., 1957; C.C.Gillispie, *The Edge of Objectivity*, Princeton, 1960. C.Raven, *John Ray, Naturalist*, Cambridge, 1942 is still the standard life. On earlier natural history, see T.H.White, *The Book of Beasts*, London, 1954; B.Rowland, *Animals with Human Faces*, London, 1974; J.W.Krutch, *Herbal*, Oxford, 1976; and B.Henrey, *British Botanical and Horticultural Literature before 1800*, 3 vols, Oxford, 1975, a magnificent and exhaustive survey. On the emergence of scientific societies and of science as a career, see my *The Nature of Science*, London, 1976; and for the impact of printing, see E.L.Eisenstein, *The Printing Press as an Agent of Change*, 2 vols., Cambridge, 1979.
3. On American naturalists, see J.Kastner, *A World of Naturalists*, London, 1977, though he underestimates the theoretical interest of classification; and M.R.Norelli, *American Wildlife Painting*, New York, 1975. On Linnaeus, see F.Stafleu, *Linnaeus and the Linnaeans*, Utrecht, 1971; W.Blunt, *The Compleat Naturalist*, London, 1971; W.T.Stearn's introductions to the reprints by the Ray Society of Linnaeus, *Species Plantarum* (2 vols., 1957–9), and his *Flora Anglica*, with Ray's *Synopsis Methodica Stirpium Britannicarum* (1973); W.T.Stearn, *Botanical Latin*, Newton Abbot, 1966; and G.H.M.Lawrence, *Adanson*, 2 vols., Pittsburgh, 1963. On Mme Merian, M.S.Merian, *The Wondrous Transformations of Caterpillars*, London, 1978. On illustration, see W.Blunt, *The Art*

of Botanical Illustration, London, 1950, and my *Zoological Illustration*, Folkestone, 1977.

4. A.O.Lovejoy, *The Great Chain of Being*, Cambridge, Mass., 1936; R.W.Burckhardt, *The Spirit of System: Lamarck and evolutionary biology*, Cambridge, Mass., 1977; W.Coleman, *Georges Cuvier, Zoologist*, Cambridge, Mass., 1964; and on palaeontology, the fascinating work of M.J.S.Rudwick, *The Meaning of Fossils*, London, 1972. See also R.Porter, *The Making of Geology, 1660–1815*, Cambridge, 1977, and R.Porter and L.J.Jordanova (ed.), *Images of the Earth*, Chalfont St Giles, 1979. On different national traditions, see M.P.Crosland (ed.), *The Emergence of Science in Western Europe*, London, 1975; and on British naturalists and institutions, D.E.Allen, *The Naturalist in Britain*, London, 1976. On classifying invertebrates, see M.Winsor, *Starfish, Jellyfish and the Order of Life*, New Haven, 1976. D.King-Hele (ed.), *Essential Writings of Erasmus Darwin*, London, 1968 is a useful collection.

5. H.B.Nisbet, *Goethe and the Scientific Tradition*, London, 1972; G.D.Brittan, *Kant's Theory of Science*, Princeton, 1978, are useful accounts of German science at the opening of the nineteenth century. L.Agassiz, *Essay on Classification*, ed. and intr. E.Lurie, Cambridge, Mass., 1962, and *Studies on Glaciers*, ed. and tr. A.V.Carozzi, New York, 1967. J.H.Brooke is writing a study of natural theology; with which chapter 3 of my *Natural Science Books in English*, London, 1972 is also concerned. On systems of classification of birds see the long introduction to A.Newton, *A Dictionary of Birds*, London, 1893–6; and A.M.Lysaght, *The Book of Birds*, London, 1975.

6. C.Clair, *A History of European Printing*, London, 1976; M.Plant, *The European Book Trade*, 2nd ed., and *The English Book Trade*, 3rd ed., London, 1965 & 1974; H.D.L.Vervliet, *The Book, through five thousand years*, London, 1972. M.J.Petry (ed. and tr.), *Hegel's Philosophy of Nature*, 3 vols., London, 1970, is a fully annotated translation. P.R.Sweet, *Wilhelm von Humboldt*, 2 vols., Columbus, 1978–9. On language, S.Potter and L.Sargent, *Pedigree: words from nature*, London, 1973.

7. C.Lyell, *Scientific Journals on the Species Question*, ed. L.G.Wilson, New Haven, 1970; C.Darwin, *Natural Selection*, ed. R.C.Stauffer, Cambridge, 1975 (the draft of the big book that in the event was compressed into the *Origin*), and *Collected Papers*, ed. P.H.Barrett, 2 vols., Chicago, 1977. H.E.Gruber, *Darwin on Man*, prints an interesting notebook with much editorial matter, some of it rather speculative; M.Allan, *Darwin and his Flowers*, London, 1977, shows him as a botanist. On the reception of Darwinism, J.R.Lucas, 'Wilberforce and Huxley: a legendary encounter', *Historical Journal*, 22 (1979) 313–30, shows how far the received view is from what must have happened; and J.R.Moore, *The Post-Darwinian Controversies*, Cambridge, 1979, follows the fate of the theory among protestants in Britain and America. See also W.H.Thorpe, *Purpose in a World of Chance*, Oxford, 1978. H.P.Moon, *Henry Walter Bates*, Leicester, 1976.

8. On modern science in its social context, see M.Berman, *Social Change and Scientific Organisation*, London, 1978, and J.R.Ravetz, *Scientific Knowledge and its Social Problems*, London, 1971. Insiders' views of science, or anyway physics, are in J.Bernstein, *Experiencing Science*, London, 1979, and J.Ziman, *Reliable Knowledge*, Cambridge, 1978. See also S.J.Gould, *Ever since Darwin*, London, 1978, and *Ontogeny and Phyllogeny*, Cambridge, Mass., 1977. Attractive books with discussions of classification are S.D.Ripley, *Rails*, Boston, 1977; J.Hancock and H.Elliott, *Herons of the World*, London, 1978; D.Goodwin, *Crows of the World*, London, 1976, and *Pigeons and Doves of the World*, 2nd ed., Ithaca, 1977. On recent work on man's origins, see R.E.Leakey and R.Lewin, *Origins*, London, 1977; and on biology and ethics, the essay in T.Nagel, *Mortal Questions*, Cambridge, 1979. For recent ideas, see V.H.Heywood, *Plant Taxonomy*, 2nd ed., London, 1976. On coelacanths, J.L.B.Smith, *Old Fourlegs*, London, 1955; and for physics as a dance, G.Zukav, *The Dancing Wu Li Masters*, London, 1979. For a new angle on ecology and the environment by a cyberneticist who sees a self-maintaining balance and urges that we need not be too pessimistic, see J.E.Lovelock, *Gaia*, Oxford, 1979. These lists do not pay all my intellectual debts, but do indicate recent works in the field, In addition, there is the multi-volume *Dictionary of Scientific Biography*, New York, 1970 still in progress (though this is weak on naturalists and often whiggish, as on Swainson), and other biographical dictionaries. For a general discussion of sources and problems in the history of science, see my *Sources for the History of Science*, Cambridge, 1975. There are societies concerned with the history of science; that in the USA publishes the journal *Isis*, and that in the UK *The British Journal for the History of Science*. These, and other general journals such as *Annals of Science*, include recent work and reviews in this field; but especially interesting is the historical *Bulletin* of the British Museum (Natural History), and the *Journal of the Society for the Bibliography of Natural History*. The various societies also organise meetings and conferences from time to time.

Index

Academies, 44, 52ff, 62, 68, 73f, 76, 86, 97, 111, 118, 122f, 139, 143, 147, 150, 163, 168f, 176ff
Accident, 21, 23
Adanson, M. 67, 79, 117, 192, 207
Affinity, 104, 108ff, 115, 119, 126ff, 136, 148, 150, 161, 185
Agassiz, L. 111ff, 129, 154, 156ff, 160, 167, 184, 208
Age of Man, 154, 158ff, 168, 172, 174ff, 180ff, 195ff, 204
Aggression, 172, 196f
Alphabet, 50f, 131f, 139, 149
Ampère, A. M. 107, 137, 140f
Analogy, 19, 39ff, 85, 101ff, 108, 115, 123, 149, 152, 173, 190, 193
Analysis, 136ff, 143
Anatomy, 53ff, 69, 77, 105, 119, 121, 123, 127ff, 172, 179, 193
Anomaly, 30, 39, 89, 189
Apes, 16, 35, 55, 57, 79f, 83, 89, 93, 101ff, 151, 167, 171ff, 178, 183, 195ff
Aristotle, 19f, 37, 40, 46, 52f, 56, 86, 101, 108f, 117, 171, 196, 204f
Artificial, 21, 23, 27, 46, 57, 58ff, 64, 77, 89, 118, 129, 139, 145, 173, 201, 206
Astronomy, 36ff, 57, 139f, 141, 166, 173, 179, 182, 187, 205, 207
Atoms, 22f, 52, 91, 107f, 132, 136f, 143, 166, 186ff

Banks, J. 52, 68, 73, 78, 96f, 118
Bates, H. W. 202f, 208
Behaviour, 48, 87, 123, 193f, 196, 198f

Bell, C. ix, 193
Bell, T. 163
Binomials, 65ff, 71f, 120, 184
Biology, 91f, 108, 111, 194, 199
Birds, 33, 46, 53ff, 65, 71, 97, 99, 102, 106, 108f, 127, 174, 194, 208f
Boerhaave, H. 58, 62, 74
Borderlines, 17f, 22f, 30ff, 63, 78, 82, 84, 173, 189
Borrowing, 154, 184ff, 190
Botany, 47, 55, 58ff, 70, 75, 78, 81, 82, 117f, 161, 169f, 179, 207
Boyle, R. 30, 44, 45, 132, 200
Brewster, D. 43
Brown, R. 118
Buckland, F. 25, 162
Buckland, W. 159, 176
Buffon, Comte de, 70, 79, 92, 129, 180

Catastrophe, 88, 112, 158f, 161, 168
Categories, 26f, 200
Cause, 21, 29f, 77, 88, 135, 155, 173
Chain of Being, 22, 27, 42, 78, 80, 83ff, 93, 103, 108, 113, 199
Chemistry, 17f, 23f, 49, 69f, 93, 107f, 123, 136ff, 140, 166, 183, 185ff, 190, 198, 200, 203
Children, J. G. 98, 124
Christianity, 33, 42, 44, 48, 51, 61, 79, 105, 114, 154, 168, 170, 175, 181, 200
Circles, 94ff, 100ff, 108f, 124ff, 161, 167
Coleridge, S. T. 143, 145
Collection, 27, 56, 62, 68, 72, 76, 87,

210

98f, 108, 117f, 120, 125, 149, 166, 193, 194, 202

Collinson, P. 72, 74, 75, 77, 190

Commitment, 68, 100, 114, 139, 167, 201

Conservatism, 118, 156f, 167, 176, 203

Consistency, 82, 100, 134, 178, 198

Convenience, 56, 66, 72, 77f, 82, 117ff, 160, 188

Convention, 17, 60, 67f, 84, 119f, 129, 138, 160, 173, 206

Cook, J. 73, 78, 87, 149, 196

Copernicus, N. 38ff, 52, 80, 200, 207

Correlation, 86f, 126, 134, 176

Creation, 90, 94, 113, 157, 161, 163, 166

Crookes, W. 32, 185ff

Cuvier, G. 25, 86ff, 91ff, 94, 100, 104f, 106, 107, 112ff, 121, 123, 126, 129, 135, 159, 161, 199, 204

Dalton, J. 30, 76, 107, 137, 187, 200

Darwin, C. 13f, 20, 22, 28, 34, 41, 49, 76, 87, 89f, 92, 94, 100, 103, 106, 111, 114, 124, 129, 154f, 158ff, 163ff, 184f, 189, 195, 200f, 203, 204

Darwin, E. 90f

Davy, H. 18f, 24f, 34, 57, 89, 91, 98, 107, 137, 185, 205

Definition, 24

De La Beche, H. 160, 162

Descartes, R. 26, 50, 141, 201

Descent, 14, 80, 90ff, 159, 166ff, 174, 180, 182, 190

Description, 17, 24, 57, 107

Details, 14, 53, 61, 64, 107, 116, 124

Development, 13, 22, 35, 69, 79, 84, 88, 92, 112, 114, 152, 158ff, 184

Diagnosis, 47, 67, 134f, 160, 191

Dictionary, 146f, 209

Disciples, 52, 89, 168f, 201

Disease, 46ff, 52, 132, 134ff, 140

Dissection, 53, 77, 189f

Distribution, 46f, 49, 56, 72, 74,

105, 113ff, 116f, 161ff, 169f, 183, 195

Divergence, 91, 152, 156, 164, 195

Diversity, 152, 155, 158, 160, 185

Divisions, 17, 40, 55f, 64, 172, 198, 206

Dogma, 21, 78, 113, 151, 154, 178, 201, 204

Ecology, 35, 113f, 125, 194, 202

Electron, 31, 187ff

Elements, 17f, 24, 31, 37f, 48ff, 134, 136ff, 140, 183, 185ff, 203

Ellis, J. 69f, 74, 76f, 83

Embryology, 110, 115, 161, 165, 172, 181f, 202f

Encyclopedias, 50f, 63, 99, 131ff, 142ff, 152, 161

Environment, 92, 113f, 115, 161, 199, 202

Essences, 17, 19f, 26, 152, 191, 202

Evolution, 33, 80, 90, 95, 110, 116, 121, 140, 154, 161, 168ff, 184ff, 195, 202

Explanation, 37, 76, 107f, 136, 153, 154f, 166, 184ff, 202

Exploration, 47, 58f, 67, 72ff, 78, 85ff, 98f, 106, 115, 118, 149, 151, 176f, 193, 202

Extinction, 85f, 88f, 92, 103, 113, 115, 148, 152, 157, 160, 164, 176, 181, 184

Faith, 21f, 51, 105, 158, 175, 181, 200, 204

Falconer, H. 176ff

Families, 19f, 40, 51, 53ff, 63, 71, 80, 101, 104, 108, 114, 137, 147, 149ff, 156ff, 182, 185, 205

Faraday, M. 107, 185

Fertility, 69, 79, 179, 184, 204

Fish, 53, 70, 102, 109, 112f, 156ff, 160, 194f, 209

Fleming, J. 89f, 94, 140

Flux, 69, 92, 121, 136, 143, 147, 152f, 179, 185, 189, 196

Fossils, 22, 25, 71, 85, 89, 103, 112ff,

156, 158ff, 165, 174, 182, 185, 194, 197, 204
Fraud, 175, 178, 205
Function, 69, 86f, 128, 180, 193
Fundamental, 19, 40, 57, 101ff, 119, 126, 139, 149, 189ff, 202

Galileo, 26, 39ff, 50, 52, 56, 140, 173, 201, 203
Gaps, 42, 64, 83ff, 105, 108, 139, 181, 194
Garden, A. 60f, 74
Genus, 17, 33ff, 51, 54, 65ff, 70ff, 79f, 108, 113, 117, 120, 136, 172, 184, 195
Geology, 22, 70, 84, 88, 112, 114f, 156ff, 167f, 176ff, 182
God, 29, 37f, 42, 49, 51, 60f, 75, 77, 83ff, 105, 111, 113, 115, 123, 129, 141, 143, 148, 167, 180, 185, 200
Goethe, W. 110f, 142, 190
Gray, A. 169f, 173, 180
Grew, N. 56, 69

Habitat, 72, 74, 92, 102, 113, 161, 180
Haller, A. 82
Hegel, G. F. W. 142, 145
Herschel, J. 99f
Herschel, W. 43, 140
History, 16, 85, 88, 135, 145, 151, 154, 159ff, 181f, 184ff, 190, 198, 203f
Homology, 19, 39f, 108ff, 121, 123, 149, 163, 182, 193
Hooke, R. 44f, 132
Hooker, J. 124, 169ff
Humboldt, A. 28, 52, 115f, 118, 150ff, 205
Humboldt, W. 151ff
Hunting, 23, 54, 122, 193ff, 196f
Hutton, J. 84
Huxley, T. H. ix, 41, 94, 101, 111, 161, 169ff, 183, 201, 203
Huygens, C. 42f
Hybrids, 33, 69, 90, 179, 184
Hypothesis, 38, 41, 49, 90, 134, 143,

151, 166f, 185

Identity, 54, 57
Illustration, 33, 47, 53, 56f, 59, 61, 64, 68, 69, 75, 78, 112f, 115f, 120, 125, 128, 156ff, 160
Indexes, 52, 67, 131ff, 146f
Induction, 20, 63, 100, 124, 143, 153, 192
Institutions, 22, 43, 44, 52ff, 61ff, 68, 73, 76, 79, 86f, 90f, 94, 98, 111f, 117ff, 122f, 124f, 137, 143, 147, 150, 160, 168ff, 176f, 205
Instruments, 56, 73, 124, 140, 185, 193
Interpretation, 29, 39, 49, 134, 149, 170ff, 175, 188, 200, 203
Invertebrates, 14, 23, 25, 50, 71f, 88, 91f, 121ff, 158, 178, 189

Jefferson, T. 50, 86, 87, 161
Jones, W. 150
Journals, 52, 68, 73, 132f, 143
Judgement, 27, 29, 38, 55, 59, 63, 100, 109, 117, 120, 126, 149, 192, 203
Jumps, 80, 83, 111f, 122, 125, 179, 183, 199
Jussieu, A. de, 79, 96, 117f, 129
Justification, 20, 23, 55, 59, 100, 120f, 139
Kalam, 33, 66
Kant, I. 26, 141ff, 200, 205
Kelvin, Lord, 22
Key, 47, 67, 82, 94, 131, 135, 136, 139, 156, 191, 194, 200
Kingdoms, 70, 84, 100, 140
Kirby, W. 123

Lamarck, J. B. 91ff, 105, 107, 108, 121ff, 152, 166f, 175, 180, 185, 199
Language, 26, 38, 40, 44ff, 50f, 62ff, 65ff, 80, 91, 107, 132f, 145ff, 148ff, 172, 196
Latreille, P. A. 121f, 124, 127
Lavoisier, A. L. 24f, 49, 74, 136, 203
Laws, 22, 31, 60, 69, 143, 153ff, 181,

200
Leakey, R. E. 195
Liebig, J. 13, 73, 111
Life Cycle, 56, 102, 121f, 125, 189
Lindley, J. 119, 129
Links, 80, 83, 122, 125, 139, 151,
    174, 178, 181
Linnaeus, C. 24, 56f, 60ff, 82, 89,
    92, 96, 117f, 120, 122, 126, 129,
    139, 151, 167, 171, 184, 190, 192,
    201, 204
Lockyer, N. 140, 187
Lumping, 65, 116f, 120, 124, 181,
    184, 195
Lyell, C. 88, 91, 103, 114, 158, 160f,
    167f, 175ff, 195

Macgillivray, W. 128f
MacLeay, W. S. 93ff, 106, 115,
    122ff, 126
Magic, 26
Malpighi, M. 56, 69
Mammals, 72, 84, 86, 97, 103, 108,
    110, 158, 168, 174
Mankind, 15f, 19, 27, 42, 46f, 50, 57,
    72, 79f, 89, 93, 95, 97, 101f, 110,
    114f, 129, 135, 141, 148ff, 166ff,
    193ff
Marsupials, 104f, 158, 163, 172
Mathematics, 17, 20, 37ff, 70, 80,
    95, 97, 107, 129, 141, 145, 159,
    166, 170, 206
Maxwell, J. C. 31, 37, 107
Mechanism, ix, 26, 29, 69, 77, 107,
    141f, 155, 185, 192
Medicine, 46ff, 52, 58f, 62, 118, 132,
    134ff, 151, 180, 205
Mendeleev, D. I. 138f, 194
Merian, M. S. 65, 207
Metaphor, 20, 108, 110, 185, 198
Method, 21, 117, 201
Metropolitan, 58, 76, 80, 86, 92, 98,
    120
Miller, H. 158
Mineralogy, 70, 129, 143
Mistakes, 123, 145, 150, 161, 166f,
    171, 175, 176, 178, 205

Model, 37, 185, 188
Modification, 69, 90ff, 125f, 152,
    163, 202
Morality, 18f, 171, 180ff, 204
Morphology, 69, 76, 104, 123,
    127ff, 163, 179, 190
Müller, M. 153f
Murchison, R. I. 97, 159f, 177
Museums, 68, 86f, 91, 98, 106, 111f,
    117, 118f, 124, 127, 137
Mutis, J. C. 75f, 120

Naming, 13, 24ff, 54, 60ff, 65ff, 72,
    74, 78, 84, 117, 119ff, 128, 132,
    134, 184f, 195f
National Styles, 25f, 33ff, 42ff, 52,
    62f, 86f, 92, 93ff, 105f, 111, 121,
    124, 146f, 149, 178, 203f
Natural, 21, 23, 26, 27, 35, 37, 40,
    44, 48, 50, 56f, 63, 77, 79f, 89, 94f,
    101, 106, 117f, 123, 126, 129, 131,
    139, 165, 168, 173, 201, 206
Natur Philosophie, 105, 110f, 114,
    121, 141f
Newton, I. 22, 26, 41, 45, 52, 63,
    141f, 200, 205
Novelty, 58, 116, 124, 127, 128, 161,
    171, 203
Numbers, 17, 34, 63f, 82, 89, 95ff,
    100f, 104, 124, 137ff, 187, 192ff

Obsolescence, 15, 141, 142f, 170,
    172, 184, 189
Oken, L. 105, 111, 115
Ontogeny, 179f, 182, 202f
Originality, 63, 76, 124, 127
Owen, R. 87, 111f, 154, 157, 167ff,
    176

Paley, W. 141, 143, 170, 181, 204
Paradigms, 15, 34, 41, 50, 83, 106,
    107, 142, 167, 184, 201, 203f
Parasites, 122, 126, 127
Patronage, 53, 62, 69, 73, 75, 90,
    97ff, 112, 116, 132, 150
Pepys, S. 53

Pattern, 79, 82ff, 115, 156ff, 167
Persuasion, 167ff, 175
Pests, 23, 122, 206
Phrenology, 150, 190
Physical Science, 16ff, 20, 30f, 36ff, 107, 139, 140, 143, 155, 166, 184ff, 190, 200, 203
Physiology, 77, 119, 127f, 170
Planets, 36ff, 44, 201
Plato, 17, 19, 37, 77, 113, 168
Populations, 79, 113f, 116, 122, 160, 184, 196, 204
Prediction, 26, 37f, 41, 49, 86, 103, 138f, 188f, 194, 198
Prejudice, 151, 170, 172, 175, 184, 205
Prichard, J. C. 151
Primitive, 29, 33, 80, 95, 174, 181, 196
Principles, 14, 21, 27, 100
Professionalism, 13, 32, 40, 44, 58ff, 73ff, 76, 89, 97f, 106, 107, 112, 117f, 122f, 124, 140, 166, 170f, 176f, 185, 192, 201
Progress, 28, 85, 91f, 115, 156ff, 161, 175, 178, 180, 185, 201
Prout, W. 185
Provincial, 61f, 76f
Provisional, 176, 184, 192, 200, 205f
Psychical Research, 32f
Publication, 52f, 62, 72f, 75, 79, 89, 91f, 97, 116, 124f, 132, 145, 153, 160, 163, 166, 171, 177
Purpose, 23f, 49, 86f, 131, 180, 193, 199f, 206

Quadrupeds, 25, 57, 71, 72, 86, 100
Quarrels, 30, 89, 98f, 111, 126, 160, 166ff, 201

Race, 79f, 95, 114f, 120, 173f, 178ff, 199
Rank, 42, 57, 80, 83f, 88f, 95, 110, 127ff, 140ff, 145, 154, 158, 166, 173, 179ff, 192, 196, 198ff
Ray, J. 46, 52ff, 58, 60, 77, 127, 163,

Realism, 17ff, 21, 113ff, 117, 119, 173, 201, 206
Reason, 26, 43f, 101, 110, 167f, 171, 178, 180, 193, 206
Reconstruction, 86, 103, 194f, 197
Reduction, 141f, 184ff, 190f, 198
Relationships, 14, 19ff, 29, 39f, 53f, 80, 100ff, 109f, 113ff, 119, 138, 149ff, 156ff, 202
Relevance, 27, 56, 104, 167, 192
Renaissance, 25, 35, 36, 51, 54, 102, 105, 119, 159f
Representation, 102
Reptiles, 25, 71f, 102, 112, 157, 163
Responsibility, 170f
Rutherford, E. 187f

Sciences, 30, 74, 107, 131, 140ff, 170
Sedgwick, A. 98, 159f
Selection, 155, 163, 168, 181f, 185, 193, 195, 200, 202, 205
Semi-conductors, 31
Sex, 56, 63f, 69, 91, 117, 155, 198
Slavery, 79, 150f, 170, 181
Smith, J. 68, 78, 81, 96, 118f
Smith, J. L. B. 194f
Society, 27ff, 33ff, 45, 73, 80, 84f, 91, 94, 98, 102, 117, 135, 150ff, 160, 168ff, 175, 179ff, 203ff
Solander, D. C. 73, 78
Spencer, H. 28
Species, 19, 33ff, 47, 55, 65ff, 69, 71ff, 79, 84, 88, 95, 113ff, 120, 129, 136, 151, 156ff, 163, 174, 178f, 184, 195
Specimens, 54, 60, 69, 74, 115, 120, 128
Speculation, 24, 38, 79, 90, 92, 95, 103, 108, 123, 151, 185
Splitting, 65, 116f, 120, 124, 135, 159, 184, 195
Stability, 65, 67, 69, 93, 108, 111, 114, 119, 134f, 185, 199
Statistics, 159, 167
Stephens, J. F. 124f
Stokes, G. 22, 107
Stratigraphy, 159f, 177, 183

Struggle, 28, 153

Sub-species, 68, 121, 129, 184

Superficial, 19, 40, 61, 69, 76, 101ff, 119, 128, 135, 149, 170, 189, 202

Swainson, W. 93ff, 97ff, 107ff, 115, 124, 127, 189, 203

Symbols, 34, 45, 48

Synonyms, 54, 65, 119

System, ix, 21, 23, 33f, 46, 48, 50, 54f, 58ff, 78f, 82, 91f, 103, 113ff, 124, 126, 136, 206

Tables, 46, 62f, 70f, 136ff, 185ff, 194

Teaching, 52, 55, 58ff, 64, 112, 119, 125, 135, 145, 151, 153, 205

Technology, 22, 31, 56, 146, 181

Textbooks, 161, 177

Theory, 22, 25, 28, 31, 41, 56, 76, 90, 106, 108, 121, 126, 128, 134ff, 139, 143, 167, 181, 200

Thompson, J. V. 14, 125

Thomson, J. J. 31f, 187f

Time, 69, 88, 90, 113, 135, 141f, 154, 156ff, 163ff, 169, 175ff, 182, 184, 195, 205

Topsell, E. 50, 131ff

Tournefort, J. P. 56, 58, 60

Translation, 44f, 149ff

Truth, 17, 22, 38ff, 41, 77f, 103, 106, 126, 167, 184f, 192

Type, 54, 60, 68, 110f, 115f, 139, 163

Tyson, E. 55, 172

Uniformity, 36ff, 82, 100, 104, 110f, 134, 152, 158, 167, 172f, 185, 200

Variation, 60, 68, 82f, 115, 120, 134, 154, 181, 184

Vestiges, 159, 161, 165ff, 183, 198

Vitalism, 89, 111, 141f

Wallace, A. R. 28, 34, 161ff, 173, 202

Waterton, C. 193

Weight, 65, 79, 89, 101, 104, 117, 127, 129, 137, 143, 182, 186, 189, 192, 202

Westwood, J. O. 125

Whewell, W. 43, 133, 143ff

Wilberforce, S. 168ff

Wilkins, J. 44ff, 53, 66

Willughby, F. 46, 53f, 127

Women, 75f

World-view, 15, 34, 41ff, 50, 82f, 88, 100, 107, 113, 141ff, 161, 166, 184ff, 198

Zoology, 47, 55, 70ff, 81, 82, 89, 97, 106, 117f, 120ff, 140ff, 161, 173, 179, 189

Zoos, 86, 90, 193